OPPORTUNITIES FOR SCIENCE IN
THE PRIMARY SCHOOL

OPPORTUNITIES FOR SCIENCE IN THE PRIMARY SCHOOL

Alan Peacock

Trentham Books

First published in 1997 by Trentham Books Limited

Trentham Books Limited
Westview House
734 London Road
Oakhill
Stoke on Trent
Staffordshire
England ST4 5NP

British Cataloguing in Publication Data
A catalogue record for this book is available from the British Library
ISBN: 1 85856 017 9

Acknowledgements
Sue Eland, Leicestershire LEA, for fig. 1
Liverpool University Press for figs. 2 and 3 (SPACE Research Reports: Light, p.26, Growth p.26)
Macmillan Kenya for fig. 4 (Beginning Science, Standard 7, Alex Berluti, p.28)
Pergamon Press for fig.5 (Issues in Science Education, Husen and Keeves (eds), pp.163-164)
Software Production Enterprises for fig. 6 (Making Sense of Science, John Stringer/Channel 4 Schools, p.35)
Thos. Nelson and Sons for fig. 7 (Bath Science 5-16, Key Stage 2B, Exploring Materials p. 11)
Nick Pratt, University of Plymouth, for figs. 8 and 9
Redhills Combined School, Exeter, for figs. 12-14
Geoff Peacock, Peak Graphics, for figs. 15, 22-25
Royal College of Physicians for fig. 16
Vanessa Francis, Maskew-Miller Longman, Johannesburg, for fig. 26
Longhorn Kenya for figs. 27 and 28 (Start Finding Out Standard 4, Beryl Kendall, p. 36, 39)
Handspring Trust, Johannesburg, for fig. 29 (Spider's Place, How to Become a Great Detective, p. 17)
Collins Educational for fig. 30 (Nuffield Primary Science Key Stage 2, Electricity Pupils' Book pp.2-3)
Birmingham Development Education Centre for fig. 31 (Why on Earth? Barnfield et al., p.11)
Viking/Penguin for fig. 32 (Andy Goldsworthy).

Designed and typeset by Trentham Print Design Ltd., Chester
and printed in Great Britain by Bemrose Shafron (Printers) Ltd, Chester

Contents

TO NICK AND LINDA

CHAPTER 1

THE EMERGENCE OF PRIMARY SCIENCE

The origins of Primary Science

As a young teacher in the mid-1990s, 'Primary Science' is a pheno
menon of your lifetime. Until the fifties, the term 'science' was hardly
ever used in primary schools; in my small-town primary school in
Yorkshire we had 'Nature Study', and went for nature walks to look at
trees and flowers, and to collect frogspawn; back in the classroom we
would draw the developing tadpole, keep weather records, make a
display of fruits at harvest time. In grammar school, we did Chemistry
and Physics – (no Biology: it was a boys' school!) – by watching the
teacher demonstrate experiments, starting in week 1 with 'the parts of
the bunsen burner'. The secondary modern schools did a watered-
down version of this called General Science or Rural Science.

In the early sixties, however, like everything else, science curricula in
Britain began to change. In the USA, anxiety about the irrelevance of
school science to the real world, and about their falling behind the
Russians in the space race, led to massive resources being committed
to developing new science programmes for high school children. In
Britain we followed suit with the Nuffield Science materials, which
pioneered a new approach to science teaching, where pupils actually
undertook investigations themselves, rather than simply doing
practicals to verify the results the teacher had written on the board.
Prime Minister Wilson threw his weight behind these changes, stress-
ing the importance of effective science education if Britain's economy

was to be revitalised and new industries forged in the 'white heat of the technological revolution'.

It took longer for these ideas to trickle down as far as primary schooling, however. Nuffield Junior Science developed from its senior siblings, but the first major impact on the primary school curriculum in Britain was made by the Schools' Council, a body made up of teachers and educationists from across the whole spectrum of education, charged with reviewing and developing new approaches to the curriculum and examinations. In the early seventies, the Schools' Council published 'Science 5-13' an approach to science for primary and middle schools which adopted the then fashionable Objectives Approach which had been the basis of most American developments in science curricula. The key volume in the materials, entitled 'With Objectives in Mind' set out the philosophy of the programme, and established for the first time a discussion about what science in primary schools ought to be trying to achieve. The objectives of Science 5-13 were on the processes of science rather than teaching facts, as Wynne Harlen, the major author of the project stated:

'...to observe, raise questions, propose enquiries, to answer questions, experiment or investigate, find patterns in observation, reason systematically and logically, communicate findings and apply learning.' (Harlen, 1978)

The approach was consistent with the child-centred approach at the heart of 'good primary practice' embodied in the influential Plowden Report of the late sixties and was widely popular; but despite this, it was never used by more than a quarter of primary schools. It marked the first attempt in Britain to foster the teaching of science as a process to primary age children; some commentators such as Paul Black have suggested that for this reason, teachers lacked the confidence to take up the new philosophy. However, the project's materials have enjoyed a remarkably long shelf-life, and many of today's commonplace notions in primary science (such as Minibeasts) had their origins in Science 5-13. The books are probably still on the shelves in your school after over 20 years, as they are in many other schools; have a look!

Rumblings of discontent

This takes us on to the situation in the late seventies, when you were probably beginning primary school as a pupil. Up until then, there had been no requirement that primary teachers must teach science; the primary curriculum was largely school – and teacher-determined, and as a consequence science was still a Cinderella subject. In 1976, Prime Minister Callaghan instigated what has come to be called the 'Great Debate' about education and the curriculum, and science was again thrust to the fore of curriculum concern. However, the Inspectorate report on Primary Schools in England in 1978 only served to emphasise the problems of science practice in primary schools, asserting that few schools had effective science programmes; there was lack of equipment; the teaching of science skills was often superficial or non-existent; teachers had little confidence and lacked a working knowledge of elementary science; and that in hardly any classrooms did children investigate scientifically the questions and problems which arose from their everyday experience and interests. All this at a time when the economy had been struggling for five years since the oil crisis, and when science education was seen by politicians as a crucial element in the long-term strategy for recovery.

These events inevitably provoked an increase in the amount of science on the primary curriculum; but over the next ten years, all evidence pointed to the fact that little change took place in the actual science achievement of children, particularly in the learning skills which Science 5-13 and the Inspectorate had sought to foster. For example, the Assessment of Performance Unit showed in the early eighties that most children at age 11 were not competent in science skills such as recognising patterns in observations, hypothesising and controlling variables; the DES in 1984 claimed that science in primary schools was falling far short of the effectiveness claimed for it; and the HMI inspection of 1989 found that weaknesses identified in 1978 still persisted, placing some of the blame at the door of teachers' lack of science knowledge.

Accordingly, in 1985 the DES had taken the unprecedented step of publishing a Policy Statement on 'Science for All' which can now be seen as the precursor of the Science National Curriculum, and which stated clearly not only that all primary teachers had to include some science in their teaching, but also stipulated in broad terms the science knowledge and skills which children ought to be taught in primary schools. The stage was thus set for Kenneth Baker to instruct the Science Working Group to establish the first ever prescription of what teachers must legally teach in science in an English/Welsh primary school.

The development of Primary Science in other countries
Meanwhile, what had been happening in the rest of the world? In Europe, there had been no parallel upheaval in the science curriculum for primary schools. The German Grundschule continued to teach 'Sachkunde', which was in effect local environmental studies, emphasising general knowledge, but none of the investigation skills pioneered by Science 5-13. Likewise in France, science did not figure as a subject in the primary curriculum. The first International Science Survey of attainment in 19 countries in 1970 had served to emphasise the deficiencies which had become only too apparent in Britain, and many countries realised the need for change.

The main developments came in those areas of the world most in-fluenced by American and British systems, such as the commonwealth countries and the Pacific Rim. In Africa, for example, the Africa Primary Science Programme in the seventies developed materials which were modified and used in many African countries, materials which emphasised science problem-solving through use of the same investi-gation skills described above by Harlen, and were of a very high standard. Commercial publishers in individual countries such as Kenya took this a step further by commissioning high quality textbooks for primary schools based on the newly developed syllabuses: good examples are the 'Beginning Science' series from Kenya (Berluti, 1981 onwards), and 'Primary Science for the Caribbean: a Process Approach' (Douglass and Fraser-Abder, 1984). In some ways, then, countries like Kenya, already having science as a compulsory subject in the primary school, and having excellent teachers' and pupils' materials, were moving ahead of us. The materials were strongly influenced by 'North Atlantic' curriculum developers, and had strong family resemblances to materials in the USA and Britain: it only remained for them to be effectively implemented in the classroom.

UNESCO meanwhile had also been active in the field of primary science, producing their valuable Resource Book (UNESCO, 1973). This activity was followed by a major UN report in the late seventies which began to consider the role of science education in national development.

Various authorities however began to question the (Western) assumption, prevalent up to this stage, that science was somehow 'culture-free' and that a Western science curriculum was thus appropriate in all contexts. Links began to be made between economic development, locally appropriate vocational skills and the skills of science; and for many developing countries where primary education was terminal for the great majority of children, this meant looking again at the science content of the primary curriculum. Thus the debate about the relative importance of knowledge, process skills and attitudes in science came to the fore, not only in Britain in the lead-up to the National Curriculum, but also in many developing countries. In many such countries where science was already well established, skills and attitudes useful in agriculture, mining, conservation, tourism and other aspects of the economy of developing countries were now being seen as just as important to primary leavers as factual knowledge or exam passes, hitherto the ticket to an office job.

The National Curriculum in Science and its impact on primary teachers

Back in Britain,the Education Reform Act in England and Wales gave 'Core' status to science, and the Science Working Group moved quickly to work on the prescription of content, so that from 1987 onwards science became the 'guinea pig' subject in Kenneth Baker's strategy for statutory curriculum development. Assessment strategies were simultaneously being developed by the Task Group on Assessment and Testing (TGAT) and subsequently by the Schools Examinations and Assessment Council (SEAC), so that versions of proposals came and went with alarming frequency. Commercial publishers raced to adapt their schemes or develop new ones; more than one expensive new science package had to be abandoned or re-jigged at the eleventh hour because of changes in the number and wording of Attainment Targets.

Primary teachers waited apprehensively for the outcomes. Most of them had experienced little science teaching themselves in primary school; most were women who had rarely studied physical sciences. The big question was, would the profession be able to 'deliver' the new science curriculum? In anticipation of a 'no' answer, the DES was already making available relatively large amounts of money for in-service training, and through the new Education Support Grants and GRIST (Grant-Related In-Service Training) systems of funding in-service work. Local authorities all over the country appointed Science Advisory Teachers to run courses and provide school-based support for teachers. At the same time, the Council for the Accreditation of Teacher Education, appointed to approve all initial training courses for teachers, stipulated that trainees must receive the same minimum 100 hours training in science as in English and maths (increased to 150 hours in 1993), which precipitated changes in the way primary teachers were trained to teach science. Research such as the STAR and SPACE (Science Processes And Concept Exploration) projects were providing valuable evidence, in forms accessible to teachers, about children's science learning and ways of assessing it in the classroom.

A major effort appeared to have been made, then, to make teachers ready to teach the new science National Curriculum (although cynics might argue that the actual money available was no greater than it had been before). Has it been successful? The evidence of most evaluations so far is not encouraging. In their evaluation of the ESG initiatives, for example, the ASE evaluators found that teachers tended to revert to 'old' habits as soon as the in-service support was withdrawn (IPSE, 1988). The first evaluation of the pilot trials of Standard Attainment Tasks found that teachers themselves did not fully understand what was required, especially in open-ended investigation. And studies of teachers' science knowledge continue to show widespread misunder-standings of science phenomena, and lack of confidence in teaching the subject (e.g. Wragg 1991, Summers and Mant, 1992). Recent research is still showing that the teaching of science in primary schools is often characterised by frequent errors of fact; tight teacher control leading to aimless activity; missed opportunities to elaborate on pupils' responses; and explanations and discussions which compound pupils' own misconceptions.

At the same time, there is encouraging evidence to show that, where teachers do have background knowledge and confidence, their practice is frequently in line with accepted notions of good science practice. In other words they plan a clear framework for teaching both content and skills; have appropriate materials available; explain expectations clearly, match tasks to different abilities; listen to and challenge pupils' responses; respect their prior knowledge; and use authentic models and representations for instruction. Thus a wide range of quality in science teaching still exists in primary schools, often within the same school. How does the achievement of our children compare, therefore, with children in other countries?

Comparisons with science achievement in other countries

The most recent comparable evidence comes from the Second International Science Study of achievement in 23 countries (SISS), data for which was gathered in the mid-eighties, before the introduction of the national curriculum. At that time, just as the 'Science for All' policy statement was being issued, 10-11 year-olds in England were doing better in science than their peers in Poland, Singapore, Hong Kong, the Phillipines, Israel and Nigeria, but worse than 10-11 year olds in Japan, Korea, USA, Australia, Italy, Norway, Sweden, Canada, Finland and Hungary. Our children did better in the biological and earth sciences, worse in the physical sciences. Only about 10% of primary science was taught by teachers who had specialised in sciences; about 38% of science was devoted to practical work. Whilst we achieved relatively badly in the primary and secondary phases, our students achieved highly in the tertiary phase. Our primary school children were taught in classes averaging 28 children; in Korea, the country investing the greatest proportion of its state budget on education, primary class sizes averaged 55, and children spent 38% of their science time on practical work, exactly as in England. However, the Korean children did their science in laboratories, and were taught predominantly by men. In Australia, the only country besides England not to use a textbook for science in primary schools, and where only 5% of teaching time was devoted to science, the 10-11 year-olds nevertheless achieved better results than the English children. (Postlethwaite and Wiley, 1992).

It would be foolhardy to draw any conclusions from these two tiny selections of data from what was the largest survey of achievement ever conducted. The illustrations are provided simply to indicate that attainment in science is not related simply to one or two factors in the education of children, and to reiterate the message of this chapter that improving attainment in science in primary schools has proved to be a difficult thing to achieve. We have probably a lot to learn from experience elsewhere, and for this reason attention throughout this book will repeatedly turn to practice in other countries.

What then are the current concerns about science teaching in the primary classroom, and what political and professional issues have a bearing on these? The remainder of this chapter will try to set out some of the main conflicts – it is probably not too strong a word – which persist amongst 'primary scientists'.

The Knowledge-Process debate

An enduring concern during the past 25 years or more has been about what should be the relative emphasis on science knowledge on the one hand and science processes on the other. The Science Working Group resolved this (and four successive Secretaries of State have concurred in their creation of statutory requirements) by requiring the primary curriculum to focus equally on 'Knowledge and Understanding' on the one hand and 'Investigation' on the other. However, there has been an increasing opposition from both researchers and teachers over the years to the idea of trying to teach, or even specify, investigation skills independently of a particular context: and some authors such as Millar and Driver (1987), and Claxton (1991), would argue that this is at the root of teachers' difficulties. Researchers in other countries would also agree. For example, it has been shown in Nigeria that children's observational drawings in science are powerfully influenced by the cultural significance of the objects being drawn (Jegede and Okebukola, 1991). The specification of content in the National Curriculum originally tried to avoid separation by the use of 'Programmes of Study' which prescribe domains encompassing both the knowledge and the skills to be investigated, domains now best seen in the Level Statements of the latest version.

But more importantly, the Statements of Attainment against which children were initially assessed were specified separately for knowledge and skills; and a major lesson from research over the years has been the powerful way in which assessment methods affect what and how teachers teach. And increasingly, the economics of assessment has moved standard testing away from practical to written testing. Thus the question of the relative importance of knowledge and skills, and their integration in practice, remains at the centre of primary science practice. Politically, there has always persisted a sub-text which somewhat naively associates knowledge acquisition with 'real' learning and investigation with the 'child-centred' approaches much derided by the right; and in recent years discussion in the media has been increasingly characterised not by evidence from research (of which there has been a great deal) but by the shrill protestations of journalists of fairly extreme political persuasions. It is increasingly difficult, therefore, for the parent or interested observer to sort out professional opinion from partisan prejudice.

Assessing science in primary schools

The problem above is compounded because it is almost impossible to separate this question from the controversial issue of assessment. At present, children have to be assessed by law in science in primary schools in England and Wales (not in Scotland, interestingly, where only English and maths are assessed) and their progress reported at age 7 and 11. Assessment is by a combination of Teacher Assessment (TA) and Standard Attainment Tasks (SATs). The process has been associated with controversy from the beginning, and numerous interventions by prime ministers, secretaries of state, teachers' unions and local authorities have progressively modified the requirements and the form of the tests and reporting procedures. One contentious issue is the assessment of investigation skills, which still occupy half of the statutory requirement of teachers.

It had been convincingly argued from the beginning by TGAT and all other informed opinion that practical skills can only be satisfactorily assessed through practical means. But such standardised practical tests are both expensive and time-consuming, never mind being difficult to administer fairly (witness the 7 year-old asked to predict if the melon would float, who answered yes 'because Kate's melon floated this morning'!). In the initial trialling of practical tests with 7 year-olds, there was wide concern amongst teachers about the time taken and the consequent disruption of learning; accordingly, tests have been reduced considerably in scope, and reliance on pencil-and-paper tasks is beginning to dominate the standard tests, inevitably dominating the teaching prior to the tests.

Of course, this does not in theory prevent continuous practical assessment of skills by teachers throughout the year (although the National Curriculum as a whole has increased the amount to be taught so much that many teachers feel hard pressed to 'cover' everything). But the whole development of national curriculum requirements has been accompanied by a growing distrust of teachers by legislators. For example, the National Curriculum Council (NCC) which initially was responsible for proposing new curricula, had few teachers on working groups, unlike the former Schools Council which was required to have at least 50% of all panel members drawn from amongst teachers. The development of distrust is epitomised by the moves in the early '90s to reduce the emphasis on continuous assessment in the GCSE by John Patten. Here, it was clear that national improvements in the GCSE results in 1992 were interpreted not as an improvement in pupil performance but as a failure of examination boards to maintain their standards in the face of (implied) cheating in continuous assessment, connived at by teachers and parents alike. Thus the legislators distrust parents and teachers, whilst the teachers find that the outcomes of all this extra assessment work do not, by and large, tell them anything they didn't know already.

Thus it looks as if the assessment process is clearly not yet satisfactory; the question is, do we need it in science in the primary phase? For some, it is clearly motivated only by political desires to identify and weed out 'bad' teachers and schools, a fear accentuated by the Chief Inspector's comments in 1996 about 15,000 teachers needing to be sacked. Others in England have argued that one positive outcome has been an increase in teachers' professionalism relating to planning, assessment and reporting skills. In Australia, the decision has already been taken to abolish all forms of standard assessment in the primary phase; in some countries, all assessment at the end of primary schooling, including science assessment, is by computer-marked multiple-choice tests. In which direction are we, or ought we to be, moving?

Specialist teaching of science in primary schools

If, as this chapter has suggested, primary teachers for a variety of reasons have continued to have difficulties with teaching science effectively, is there then a case for the specialist teaching of science in primary schools? At present, only a small minority of children receive specialist teaching of science in England, although it has been commonplace in some subjects such as music for many years. Most countries in the 1984 survey did as in England, and let class teachers teach science; only in Poland and the Philippines was there extensive specialist teaching of science. At the same time, the 'Three Wise Men' report (Alexander, Rose and Woodhead, 1992) advocated moves towards specialisation at the top end of the primary school, and clear evidence has been referred to which shows that teachers with adequate subject knowledge and confidence, teach science more effectively. The postgraduate certificate route into primary teaching is still the main provider of primary teachers with a science background: yet that science background may not always be appropriate to primary school work. DfEE legislation now requires that all training courses incorporate elements relevant to the role of specialists as Curriculum Leaders in their subject area: in many schools, such specialists play a role in the professional development of their non-science colleagues. Often, they find that increased pressure on their time brought about by demands of National Curriculum legislation prevents them from fulfilling this role adequately, denying them for example time to work collaboratively with colleagues in their classrooms. As a specialist, therefore, is your expertise and time best used as a science teacher or as an in-service supporter of your class-teacher colleagues?

The goals of science education; scientific literacy or preparation for specialism?

The previous question may have to be linked to wider questions about the goals of science teaching. In the early Science National Curriculum documents, and in the Non-Statutory Guidance, attention was paid to such questions as science teaching in a multicultural society, but none of these recommendations and sentiments are enshrined in the actual legislation, which only specifies content to be taught and assessed. So as the recent OECD study emphasises (Black and Atkin, 1996) we still ask questions about whether 'society' is best served by focusing on a broad base of scientific literacy, where everyone 'speaks the language of science', or whether we ought to put our resources and effort into producing an elite group of specialists excelling in science.

In England, we have pursued the latter choice for so long that we have almost forgotten that there may be a choice; for so long, science examinations such as 'A' levels have been dominated by the narrow requirements of the few who proceed to university and scientific careers, and the trickle down of academic approaches has consistently influenced science teaching and curricula at all levels. Attempts to reverse this through such initiatives as continuous assessment at GCSE have, as we have already seen, been the victims of their own success. The British Association for the Advancement of Science, amongst many other bodies, has bemoaned the lack of public awareness and interest in science issues, and has tried to act to change things in recent years. Meanwhile, the World Conference on Education for All in 1991 came down strongly on the side of basic scientific literacy, and prestigious scientists worldwide have argued the same way at international conferences. Yet our existing National Curriculum in science focuses only marginally on those 'big issues' which people need to be informed about, such as global resources, environment, pollution, health, conservation, genetics, agriculture. Do we want all our future citizens to understand and speak the language of science? In other words, do we want a population accepting that it is good for everyone to be scientific in the way they deal with problems in everyday life, or are we happy to go on thinking that there are a few clever people who become Scientists, while most of us never will? At the moment, the evidence of what is said and what is done tends to conflict.

Effective training for primary teachers of science

The early nineties have been characterised in education by an attempt to change the nature of teacher education, and specifically to transfer more of the responsibility for this to the schools themselves, first through the Licensed Teacher and Articled Teacher schemes, then through Partnership requirements, mentoring and reallocation of funds for training in the direction of school-based training. In keeping with its dislike of the 'progressive' teaching supposedly fostered by teacher educators, the far right has gone so far as to suggest that teacher training as we know it is largely unnecessary. Certainly the Teacher Training Agency since it took over the funding of training has made damaging cuts. I once discussed this question with an experienced Kenyan teacher educator, who laughed uncontrollably at the prospect of a developed country like ours deliberately imitating the situation many developing countries seek to escape from, namely a dependence on untrained teachers. I asked him why it was a bad idea. 'Because teachers can only teach what they know!' he replied.

The question of what teachers know, or what they should know, is an important one in primary science, since it relates to many of the issues discussed above. Primary teachers by and large still have inadequate science background knowledge, on their own admission; so should training focus more intensively on this? Many would argue that it should; on the other hand, there is also evidence that the time allocated to science in initial training courses, even since the increase to a statutory 150 hours, still falls far short of the time needed to give non-scientists (the majority of primary teachers, not only in Britain but throughout the world) sufficient knowledge and confidence. Will the effects of compulsory science through the National Curriculum, now that they have worked through to those in initial training, be sufficient to rectify the situation? It is difficult at this stage to be optimistic. The work on children's perceptions in science initiated by Driver, Osborne and others, which has spawned the whole 'constructivist industry' in science has only served to show that, even with graduate physicists, childhood misconceptions about physical phenomena can persist into adult life. Moreover, Solomon (1987) for example, has reminded us that ideas are socially constructed; what we learn is influenced by those around us. Are we then destined to a continuation of the situation in which only a small minority will ever be seen as scientists?

The professional education of primary teachers in science has instead tended to focus in recent years on learning the processes or investigative skills of science, and the methods appropriate to teaching these to children. Many high-quality materials have been produced to support teachers, in many countries, and much has been learned and incorporated into such materials from research into children's perceptions, exemplified by the Nuffield Primary Science materials, based on the research of the SPACE Project. (Though there are concerns about the extent to which these materials are actually used, as will be discussed in chapter 6). Science in ITE courses has increasingly provided workshop activity which models the kinds of activity considered to be appropriate to children's learning. But it is becoming increasingly accepted that the education of effective primary science teachers is a long-term process, not merely an intensive initial experience topped up by the odd day of in-service. Even with first class mixed media resources, for example, evidence from South Africa shows that most teachers and trainers have a limited repertoire of strategies for using them, and therefore do not do so effectively (Perold and Bahr, 1993). So what form should the continuing education of primary teachers take? And controversially, what aspects of science teaching can initial training 'leave' till later in a teacher's career?

In the mid-eighties, when in-service training was under the spotlight and much evaluation was undertaken both in Britain and elsewhere, a number of lessons were learned. For example, there was a consensus that effective INSET needs to be recurrent; it needs to be located to a larger extent in teachers' definitions of their own needs; most teachers prefer INSET that is collaborative, collegial, school-based and grounded in a specific context; and specialist input or modelling of effective teaching is an important element. Science Advisory Teachers have played an important part in stimulating this kind of INSET work, in a variety of ways. However, the regressive cuts in education budgets in recent years in Britain have terminated much of this valuable work, and few Advisory Teachers remain. Under Local Management, schools have to settle for do-it-yourself INSET with less and less access to specialist input from outside. In many countries, in-service in science is in the hands of non-governmental organisations (NGOs) funded by international aid agencies or industrial conglomerates, and is therefore also at the mercy of political fashions and the goals of multinational companies. The role of the science specialist within the school thus becomes crucial; for despite the prescription of the 'what' of science by national or centralised curricula, the 'how' is still largely in the hands of individual schools and teachers, despite the impact of Ofsted. So we still have to decide how should science best be 'packaged' in the interests of the learners and the teachers. For example, do we stick to science taught through the typical cross-curricular topics, or go for specialist science topics, or even teach science as a separate subject by specialist teachers?

The relevance of primary science

One issue arising out of this is the perpetual question of relevance touched on above. For many years, the primary curriculum was seen as a preparation for work in the secondary phase, and culminated for many in the hurdle of 11-plus verbal reasoning and other such tests. In many countries throughout the world, however, science is still examined at the end of the primary phase, which is in effect the end of schooling for the majority. In the sixties, comprehensive education and the Plowden Report institutionalised notions of integration and child-centred approaches, which only a few local authorities such as Leicestershire and the West Riding of Yorkshire had previously adopted. But despite this move to start from where the child was and develop work out of each child's capacities and interests, the HMI Report of 1978 expressed anxiety about how rarely this happened in science, even though such initiatives as 'Breakthrough to Literacy', to take just one example, had been successful in individualising language work. At that time, teachers by and large took little trouble to find out what children's science interests and knowledge were, and did less to use them as starting points. We now have much clearer ideas about children's science knowledge when they begin school, and their increased interaction with mass media has also provided insights into children's science beliefs, interests and concerns. And yet we still have in primary schools a 'school science' which is in many ways only tenuously linked to the 'real world' concerns and problems which science can help them deal with. And of course in the developing world with its far more serious political, resource and teaching constraints, relevance is far more difficult to achieve (Knamiller, 1984). The turning away from science which is evident at secondary and tertiary phases, particularly amongst girls, is of great concern; recruiting sufficient high-quality scientists into teaching not only continues but is increasing as a problem., in the UK as elsewhere. In South Africa, for example, it has been estimated that only 1 in 10,000 children who begin primary school matriculate in a science subject. Would these critical situations be ameliorated if school science, particularly primary school science, were seen to be concerned with the urgent questions of environment, health, nutrition, exercise, conservation, climate, energy consumption and other questions which preoccupy professional scientists as well as many children?

For example, would it concern children to know that the average British citizen uses about 50 times more energy and 20 times more water than the average citizen of the world's poorest countries, such as Nepal? If so, what else should children know and do in science that might help them deal with this constructively? Energy and Water are both in the National Curriculum; the question for teachers is whether to focus merely on what is required for children to achieve the required levels, or whether to 'go for' science learning which has a 'realness' about it. The two are not by definition exclusive; but in practice, the former still tends to push out the latter, particularly as teachers struggle and juggle with the attainment targets of all the other national curriculum subjects as well as science. Children are not likely to develop a genuine enthusiasm and scientific curiosity if, as often happens, they are asked to copy down a diagram of 'The Water Cycle' every year. Chapter 8 returns to this question in greater detail.

Postscript

In this chapter I have tried to do two things: first, to show briefly how we got to where we are in primary science (and by implication, to show also that we still have a long way to go); and second to pose some important questions about where we are going and your role as a teacher of science in this, in a wider cultural and political context. In the process, I have glossed over the detail of such things as the National Curriculum, teaching and assessment, books and other materials, equal opportunities and other important issues. Subsequent chapters will address these in more detail. My idea is that you may try, as you progress through the book, to develop your own thinking on some of the major questions raised; and I will try, in writing it, to provide some pegs on which to hang your developing ideas. In the end, you teach what you are, in science as in everything else.

CHAPTER 2

WHAT SCIENCE DO PRIMARY SCHOOL CHILDREN NEED TO LEARN?

Interpreting 'Science For All'

'Pupils should develop an understanding that Science is a human activity, that scientific ideas change through time, and that the nature of scientific ideas and the uses to which they are put are affected by the social and cultural contexts in which they are developed'. (DES, 1988, p. 70)

The above statement, from the Secretary of State's first proposals for the Science National Curriculum, defined what was then Attainment Target 22, 'The Nature of Science'. It encompassed notions that science is what people do; that scientific knowledge is not static; and that both the content and methods of science are dependent on the world-views of people. This chapter considers the extent to which teachers, parents, children and others see science in the same way, and to what extent this view has been implemented in practice.

In their 1985 policy statement referred to in the previous chapter, the DES had defined 'Science for All' in terms of the fact that science 'permeated almost every aspect of daily life'; that 'scientific method' prepared children for 'adult and working life'; that it should foster 'curiosity and healthy scepticism, respect for the environment, the critical evaluation of evidence...'; and that it helped appreciate 'our (sic) cultural heritage and an insight into man's (sic) place in the world'. It also asserted that 'Just what knowledge of facts and principles should be taught is a matter for continual review, in the light of changes and developments in science and technology in the wider world'. (DES, 1985, pp. 2-4). The continual review has certainly come about – the National Curriculum is now in its fifth version, and the number of Attainment Targets has reduced from 22 to 4 – but whether this review has taken account of changes in 'the wider world', and in what way they reflect 'our' cultural heritage is debatable. Very rarely are terms such as '*our* cultural heritage' or 'the *wider* world' defined. What the successive reviews certainly reflect is the 'realpolitik' of education in the late eighties and early nineties, a process of change driven by economic and political pressures rather than by educational imperatives.

The quotation with which the chapter began certainly reminds us that science is about knowledge, activity and beliefs; in chapter 1 it was hinted that different cultures have different views on the relative importance of these in science for primary school children, and that within Britain the tide of emphasis on each aspect has ebbed and flowed. Wynne Harlen and others in the seventies championed the cause of 'process science'; in recent years, the emphasis on concepts has been re-asserted through such studies as the Children's Learning in Science and SPACE projects. In other countries, particularly in the developing world, vocational and traditional skills and study of the environment are being emphasised as basic needs in science learning. And scientists themselves, such as the eminent physicist David Peat, have highlighted the existence of other ways of 'Coming to Knowing' in science: for example amongst the indigenous peoples of North America, where knowledge is still valued more as a verb (like our own word 'acknowledge') than as a noun implying a set of unchanging facts. (Peat, 1995). Some authors such as Claxton (1991) have questioned all these emphases, and asserted the importance of 'real world' science concerns. But (in the market language of the nineties) how do the customers see science education?

In a survey of parents of primary school children in 1988, I found that almost a quarter of the parents still thought of science as 'Biology, Physics and Chemistry'. A fifth saw it as study of 'the natural world';

Figure 1

a sixth saw it as knowledge of facts; and another sixth as 'how things work'. Less than a sixth mentioned practical investigation, and the rest did not know (Peacock, 1989). More recent studies over the period of introduction of national curriculum science have been notable for the little change which has taken place in parents' perceptions of what science is and what their children do in science at school (Peacock and Boulton, 1991, 1995). When some parents became aware of what their children did at school, it was not unusual for them to be surprised; 'does it have to be called science?', as one parent asked. In primary schools, therefore, it is not always clear where science begins and ends; certainly some of the content areas and investigation skills which the Science National Curriculum embodies (the environment; materials; gathering observational evidence) are equally seen as a part of other subjects.

Children and learner-teachers also have clear and very similar views of science and scientists, which are also well documented in research. Scientists are universally perceived as male, bearded, bespectacled boffins with wiry hair, white coats and Germanic accents, a stereotype easily recognised in (and reinforced by) the visual media. They are not 'ordinary' people; they are frequently perceived as mixing bubbling chemicals to make explosions, tending complex electrical gadgets or dissecting prostrate bodies (none of which are as yet part of the national curriculum for Key Stage 2!). For these young people, science is definitely not 'for all'; it is a pursuit of a 'brainy' and rather eccentric minority, and is not very relevant to everyday life.

Official documentation repeatedly tries to dispel such notions; for example, the NCC's Curriculum Guidance series has considered the question of science for children with special needs, asserting clearly that 'every pupil is entitled to a broad and relevant science curriculum' and that 'all children should participate fully in learning about science and through science' by 'adapting courses, approaches and resources for pupils with special needs' (NCC, 1989). Similar sentiments are also expressed in various places about girls and ethnic minorities, and books have been written to offer practical suggestions about how to turn such aspirations into reality (Browne, 1991; Peacock, 1991; Thorp, 1991; Barnfield, 1991). However, none of these optimistic aspirations are enshrined in the National Curriculum legislation on teaching and assessing science; what we find in actuality are such practices as 7 year-olds who are second-language English speakers prevented from being assessed in their mother tongue, and schools for special needs children or schools in multilingual communities coming at the bottom of their league tables, because most of their children are only 'working to' level 1.

The rhetoric of Science for All is thus still a long way from becoming a reality. Some would argue that 'real' science is a specialist activity, and never will be or should be for all. Others might draw analogies with performing arts such as Music and Dance, where a whole spectrum of participation is possible, and where people can 'be musical' or 'appreciate' music in a host of different ways. However, what we cannot escape is the notion which governments in most countries in the world espouse, namely that science is an essential, and that it has (economic) importance 'to the pupil and to society as a whole' (DES, 1985). So what is it about science that is so important?

What children already know about science

The first and perhaps most important fact to recognise is that children already have many ideas about science when they start school. They perceive science and scientists in a certain way; and they have formed concepts about the physical phenomena which they have encountered. A great deal of research has been carried out in recent years, in many countries, to investigate these early, 'naive' perceptions of children. Much of the evidence makes amusing reading: for example, children

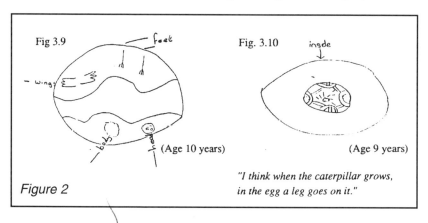

Fig 3.9   — feet
— wing

(Age 10 years)

Fig. 3.10   insde

(Age 9 years)

"I think when the caterpillar grows, in the egg a leg goes on it."

Figure 2

can draw the inside of an egg to indicate that the embryonic chick looks like a kit of parts; and many of their explanations, though misconceived, have an internal logic. For example, it is common for young children to imagine that evaporation only takes place at night (because you can't see it happening during the day, but when you come back to school next morning...).

Social constructivists have also told us that such misconceptions are common to many children, in different contexts and cultures, though without as yet exploring in any detail the cultural and linguistic contexts of learning science in different parts of the world, and how these might influence ideas. For example, it has been suggested that many children develop the idea that seeing involves an 'active eye' in such a way that something comes from the eye when you see, and that this idea is universal.

Figure 3

Equally, misconceptions about science phenomena are said to be remarkably persistent and difficult to modify. Misconceptions related to forces, for example, and such notions as 'suction' have been shown to persist well into adult life, even for graduate physicists! So the first important message about the importance of science concerns the developing of children's conceptions, and how best to help them to acquire concepts which are useful because they have more explanatory power. Social constructivists are agreed on one thing: that concept construction involves mediation by others, notably teachers and peers, and that a crucial role for the teacher of science is to assist learners by 'scaffolding' the construction process, through questioning and other forms of mediation such as providing accessible representations of the ideas to be learned. But how do we know exactly what has been learned? The final chapter returns to this difficult question.

The National Curriculum as a definition of science content for primary children

The question of which particular concepts are most important is continually debatable, and as the opening quotation reminds us, is a matter of social, cultural (and political) circumstances. The current specification in the National Curriculum has evolved, or been distilled (note the science metaphors!) from a broader specification which was considered to be impractical and unwieldy for teachers. Political influences also sought to rationalise the Attainment Targets (ATs) around traditional subject lines, in order to meet certain concerns in the secondary phase (and particularly amongst those concerned with the more able scientists who would go on to do 'A' level) about the choice between single subjects or integrated/balanced science. Thus since August 1992 we have four clear conceptual areas to be taught to primary children, dealing respectively with what at the time of writing are referred to as Experimental and Investigative Science; Life Processes and Living Things; Materials and their Properties and Physical Processes.

This specification has inevitably involved change and compromise. For example, what is left of 'Earth Sciences' after most of it was handed over to the (newer) Geography National Curriculum has been incorporated into 'Materials'; very few concepts relating to chemical change are included at the primary stage (key stages 1 and 2), in comparison with the large emphasis on physics concepts relating to heat, light, sound, electricity and magnetism; and the language of the specification still leaves large areas to be interpreted by teachers. What for example is meant by a statement which asserts that 'Pupils *know* about a range of properties, such as texture and appearance...' (AT3, level 1)? The dispute over Technology and its links with Science have also meant that Information Technology, originally a part of the Science National Curriculum, has now 'gone' to Design and Technology. And an important concept like 'Energy' has been largely located in a single Programme of Study (Physical Processes), even though it is equally relevant to Life Processes and to Materials.

These may be seen as nit-picking and unimportant arguments. On the surface, the curriculum is more or less what we would expect; it is not much different from what has always gone on in the name of science in schools in the United Kingdom, and in its gestation, the Science National Curriculum was remarkably uncontroversial, unlike some which followed such as English and History. Perhaps this reflects actual levels of popular and political significance. But how does it compare to other ideas about science? Are there alternative prescriptions which might merit consideration?

What our National Curriculum specification does, in the main, is retain the notion of science as a separate, specialist activity. But not all cultures approach science for primary school children in this way. For example, this extract from a Kenyan primary textbook indicates how science learning is tied in to traditional community activities, and deliberately emphasises the overlap between science and other areas of expertise in arts and crafts.

We have also chosen to retain more or less intact the Biology, Physics and Chemistry classification of content; but this is not necessarily the best way to make science 'important to pupils and society as a whole'. For example, many countries identify Agriculture and Health Education as subjects in primary schools, thus teaching most of the life science concepts and skills in relation to these practical, functional areas of everyday work and living. Others have proposed that Ecology or what has been called Natural Economy would be more relevant ways of organising the ideas and concerns most central to teachers and young people at the end of the twentieth century.

In many countries, the primary curriculum already incorporates Environmental Education *instead* of Science, conveying a quite different message about priorities. In 1986, for example, the National Institute of Educational Research in Tokyo published a survey of the primary science curricula in 16 Asian and Pacific countries, which between them account for about two-thirds of the world's population (NIER, 1986). The survey showed that about half the countries taught science as a specialist subject at some stage in the primary phase,

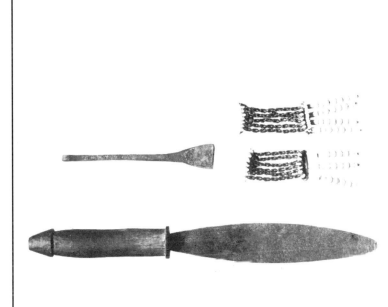

Figure 4

A knife, a razor and ear ornaments made by our ancestors using traditional technology.

Activity 2   Finding out about traditional technology

Some of our ancestors were very clever people. They did not go to school, but they learnt a great deal.

In some ways they were scientists.
They observed the properties of clay.
They discovered how to change its properties.

In some ways they were artists.
They made designs for containers.
They made patterns to decorate the containers.

In some ways they were craftsmen.
They used their hands skilfully.

Your ancestors learned how to use science, art and skill together. In other words, they invented **technology**. They used their technology to make things. These things were often very beautiful. They made tools and weapons, baskets and jewelry. Traditional technology is part of Kenya's history.

Find out as much as you can about traditional methods of making things. Ask your parents and local elders. You may be able to find out how some of the following used to be made:

soap   iron   dyes
alcohol   medicines   fuels

Try to find answers to these questions during your research:
• What things are still made by traditional methods?
• What things used to be made by traditional methods, but are no longer?
• What raw materials were used?
• What methods were used?
• Was the work done by special people? If so, why?
• Were the methods kept secret? If so, why?

whilst the other half integrated science with environmental, health and social studies. In general, they found that the 'special emphases' of science curricula in primary schools were represented as follows:

- the local environment

- a conservation ethic

- a perspective on science in relation to religion

- the preparation of a 'creative and skilled workforce'

- the integration of science into 'healthy living attitudes'

The report also emphasised the wish of most countries to move towards some idea of a common core of content for science at the primary phase. More recently, the Second International Science Survey also surveyed the emphasis in primary science curricula in 23 countries , producing statistics comparable with those of the NIER survey (Rosier and Keeves, 1991). An analysis of the two sets of results provides the basis for the 'Global Common Core' below, this being constructed from the topics most frequently emphasised in both surveys, in order of the degree of emphasis. (Peacock, 1993)

A 'GLOBAL COMMON CORE ' for PRIMARY SCIENCE
Natural Environment
(weather; cycles in nature; solar system; animals and plants; diversity; habitats; conservation; erosion; pollution; earth and rocks)

Measurement
(measuring weight; temperature)

Physical change
(change of state; dissolving; separation by evaporation; absorption of water)

Human Biology
(nutrition; metabolism; structure; illness and diseases;)

Electricity
(conductors; circuits; batteries and bulbs; magnets)

Forces
(machines)

Light
(mirrors; lenses; shadows)

Sound
(thread telephone; musical instruments)

Thus apart from the emphasis on environmental science, there is the suggestion of a mismatch between the emphasis within the specific content of primary science curricula in these countries, and their general aspirations for primary science.

But in case we should think that things are any different in Britain, we need to go back yet again to the 1985 policy statement on science. The DES was in no doubt about our priorities:

> 'In an advanced society the sheer rapidity of technological change requires high quality scientific, technical and engineering manpower on an adequate scale. It is an essential function of science education up to the age of 16 to *lay the necessary foundations* and to offer appropriate challenges for those, including the most able, who will proceed to further education and training in science and technology *and may go on to make personal contributions to the nation's scientific effort.*' (DES, 1985 p.3; my italics).

This in the fourth paragraph of the section on 'Science for All', a paragraph which ends by asserting that the proposition is applicable to the whole population. The inference is that, in our 'advanced' cultural, economic and political context, 'Science for All' in practice should be science which lays the foundations for later development of specialisms. This is indeed very different from the sentiments in the NCC document on science for children with special needs; and from notions of universal scientific literacy, described by Harari (1993) in terms such as 'science as a tool for organised thinking...for deciding what evidence can be trusted...for making sense of phenomena...for knowing what you need to know... to remove gullibility'. In Harari's terms, what the majority of people do *not* need is 'the first few chapters of a book for specialist scientists'.

There is a danger in such a debate of coming to the assumption that advanced societies 'need' one kind of science education, whilst 'developing' countries need something else. Equally, there is a danger of assuming that science has all the answers. In his work on teaching young people about the global impact of science and technology on the environment, Baez uses not only the language of ecology, but also terms such as Rights and Ethics; Peace and the Environment; the 'Special Role of Compassion' (Baez, 1991). John Lewis, a scientist and teacher, has also reminded us that children can easily come to feel that 'physics has nothing to do with life', and gives us examples of how children can be given a 'feel' for such things as units of energy, which we encounter daily on food labels but rarely relate to 'Energy' in school science (Lewis, 1991). He also provides us with this salutary reminder of the reality of our energy consumption:

'The total consumption of primary fuels in Germany is about 2 10^{19} joules. The population is about 58 10^6. So the annual consumption per person is 3.4 10^{11} joules. The average per person per day is therefore 3.4 10^{11}/365, of about 900 megajoules. But the day's work by a slave is 3 megajoules. So in Germany, each person (man, woman and child) has the equivalent of 300 slaves working for them.

That is not all. A unit of electricity, a kilowatt-hour, is 3.6 megajoules and costs about 15 pfennigs. At the same rate, 3 megajoules of work in a day would get 12 pfennigs. Thus in Germany, each person has the equivalent of 300 slaves working for them for whom payment would be 12 pfennigs per day. Is it any wonder that we in Western Europe get richer, whilst those in developing countries, who lack these 'slaves', do not?' (Lewis, 1991 p. 158).

Russell Stannard (1994) has shown that the classical Newtonian concepts of physics which we encourage our children to construct are in many ways misconceptions; moreover, in his 'Uncle Albert' books, he has shown that children can grapple enthusiastically with the amazing ideas of relativity and quantum theory, even though these are off-limits for the primary National Curriculum.

So what are the important things to study in science? Marx (1991) has argued that the lexicon of topics (of which the NIER list above is an example) does not adequately represent what it is about science that is important to a lawyer, businesswoman or politician, for example. He argues for a representation of science in terms of 'deeper ideas', and suggests a list like the following (see Figure 5).

If this looks like a daunting list, it is worth reminding ourselves of what Ausubel and Bruner, amongst others, have said about the learning of concepts and the importance of structure. The basic ideas that lie at the

Figure 5

Traditional representation of the important elements of scientific knowledge

Names of fruits, household animals, parts of the body, chemical elements, compounds;

Taxonomies of plants and animals;

Descriptive structures of cells and organs;

Laws of Hooke, Coulomb, Ohm, Kirchoff, Lenz, Boyle, Mariotte, Gay-Lussac, Curie;

Subject structures of Chemistry, Physics, Biology, Environmental Studies.

Alternative representation of the really important ideas of science

Observation, measurement, experiment, frame of reference, relativity, environment;

Composition, structure, property, function;

Reproduction, heredity, mutation, diversity, fitting, selection, evolution;

Inertia, interaction, change, anticipation, technology, handling energy and information, feedback, stabilisation, amplification;

Loss/efficiency, order/disorder, organisation/decay/pollution;

Object/subject, testing of models, complementary models;

Risk/decision, growth/stability, responsibility/future.

(*from Husen and Keeves, pp.163-164*)

heart of all science, and the basic themes that give form to life are as simple as they are powerful.

'To be in command of these basic ideas, to use them effectively, requires a continual deepening of one's understanding of them that comes from learning to use them in progressively more complex forms' (Bruner, 1960).

Thus learning 'structures for learning' in science, or meta-cognition, is central to many current ideas about how and what children learn. This means learning to 'make sense' of ideas, how they relate to each other, and how fundamental concepts (like kinetic theory, for example) have huge and liberating explanatory power, once we grasp their meaning at a level appropriate to our stage of conceptual development. In Ryle's terms, it is about 'knowing how' rather than simply 'knowing that'.

You may be thinking I have forgotten about the Science National Curriculum: not at all! My question at this point is, how does it match up to the considerations raised in the above paragraphs? How many out of ten do you give it? But before you can formulate an answer, it might help to look closely at what the Science National Curriculum now says, and what it does not say.

What the Science National Curriculum requires of teachers

The Statutory Requirements are set out in the Education (National Curriculum) (Attainment Targets and Programmes of Study in Science) Order, 1991 which, as the title of the legislation implies, requires teachers to teach science in accordance with the the specifications of **Programmes of Study** and **Attainment Targets,** at the appropriate **levels** (levels 1-3 for **key stage** 1 children; levels 2-5 for key stage 2). (In the early days of the National Curriculum, there were numerous apocryphal stories of headteachers sending back the ring binder with a letter to the effect that they had read the document and decided to stick to their own curriculum; but as yet there has been no attempt to test compliance in the courts.)

Programmes of Study represent the four main content areas which have to be taught to children. Each of the four Programmes of Study (Experimental and Investigative Science; Life Processes and Living

Things; Materials and their Properties; Physical Processes) is divided up into several strands, to make them less unwieldy: for example at Key Stage 2, Physical Processes is divided into four strands (Electricity; Forces and Motion; Light and Sound; The Earth and Beyond). These strands describe in broad terms the domains or areas of content that must be taught. For example, the first paragraph of Physical Processes for Key Stage 2 (Electricity) states that pupils should be taught:

a. that a complete circuit, including a battery or power supply, is needed to make electrical devices work;

b. how switches can be used to control electrical devices;

c. ways of varying the current in a circuit to make bulbs brighter or dimmer;

d. how to represent series circuits by drawings and diagrams, and how to construct series circuits on the basis of drawings and diagrams.

Note that this specification of content incorporates two kinds of requirements, namely the knowledge and concepts to be learned (circuit, battery, switch etc); and the investigation skills to be used, described fairly precisely by verbs such as 'construct' and 'represent'. But importantly, statements such as this do not attempt to isolate the process of investigation from the knowledge focus of children's learning. In other words, skills are intended to be learned in a specific context, and concepts are meant to be developed through the use of a range of skills. This is the intention behind the organisation of content in terms of Programmes of Study, which in turn are intended as the templates for teachers' planning of science work.

The **Attainment Targets** prescribe what pupils should know at different levels of attainment, and these two are organised according to the four programmes of study. For example, in relation to Physical Processes, at level 3 it is prescribed that:

Pupils use their knowledge and understanding to link cause and effect in simple explanations of physical phenomena, such as a bulb failing to light because of a break in an electrical circuit, or the direction or speed of movement of an object changing because of a force applied to it. They begin to make simple generalisations about physical phenomena, such as explaining that sounds they hear become fainter the further they are from the source.

Notice that, to be at level 3, pupils are required to be able to 'use knowledge' to 'link cause and effect', and 'make simple generalisations' by 'explaining'. These are the cognitive demands of level 3 in respect of this content area. Pupils are also required, in the PoS concerned with Experimental and Investigative Science, to be able to 'respond to suggestions... put forward their own... make simple predictions... make relevant observations... measure quantities... carry out a fair test... record their observations... (and) say what they have found out'.

- Thus the Programmes of Study (which you are required to teach by law) specify what to teach and the Attainment Targets (also a legal obligation) specify the criteria by which to measure the level of a pupil's attainment. However, the National Curriculum imposes no requirements on *how* you teach; there is no compulsion to use practical methods, for instance, or even demonstrations. You must *teach* pupils how to represent circuits in diagrams; but you may do it anyway you wish, and you do not have to assess that specific item of knowledge.

Ongoing Teacher Assessment of attainment is carried out in your way, according to your own teaching and methods of assessment. Only at the end of each key stage will children be formally tested by externally-set Standard Assessment Tasks (SATs). These SATs *may* require your children to plan and record construction details of a circuit by drawing diagrams; however, they will only sample a small number of skill and concept areas. The probability is, then, that they will not be required to draw any circuits they have constructed as a means to assessment unless you choose to ask them to do this. There is thus an in-built separation between what you are required to teach, and what you are required to do by way of assessment. The implications of this for the teaching of science are considerable, and you may not need to speculate long on what these are likely to be. However, we will return to this point later in the chapter.

- Science in primary schools is covered by the first two **key stages** of the national curriculum, key stage 1 (5-7 year-olds; years 1-2) and key stage 2 (7-11 year-olds; years 3-6), with Standard Assessment Tasks being set in the summer term at the end of each key stage. This is the case even where it does not coincide with transfer to secondary school; for example in authorities where transfer is at 10, the KS2 SATs are taken at the end of the first year in high school, whilst in counties with 8-12 or 9-13 middle schools, the KS2 SATs are still taken at the end of year six, one or two years before transfer.

Assessment operates by combining the teacher assessment results with the SAT results for each attainment target and producing a level for that AT. AT levels are then combined to produced an overall level for the subject, in this case science. This is done in ways prescribed by the statutory orders, the details of which need not concern us here. A report is sent to each parent indicating the levels their child has achieved, and the school and authority is required by law to publish the results of assessment at key stage 2. These results are the basis of the 'League Tables' of schools first published controversially for Key Stage 3 in 1992. The political justification for this has always been that it allows parents to make a more informed choice of school for their children; and by doing so, consumer demand will sort out the success-ful schools from the unsuccessful, thus pressurising the latter to effect improvements. Opponents of the scheme would point out that success or otherwise in the league is governed by many other factors over which schools have no control, and that it is consequently pernicious to place the entire burden of improvement on them. Others draw attention to the 'value-added' or the improvement achieved in relation to the baseline from which pupils started. Equally, it is often pointed out that it is the schools, not the parents which make the choices about which pupils go where, and that the system has not improved parental choice. This is to some extent borne out by the evidence from league tables in many authorities, where the fee-paying schools are at the top of the league, and the special schools at the bottom. Do parents of children with special needs, or those who live in rural areas, really have a choice?

The National Curriculum orders specify that levels 1-3 of the science National Curriculum are 'applicable' to children in KS1, whilst levels 2-5 apply to KS2. There are 10 levels covering the entire span of compulsory schooling from 5-16; this implies that progression through the levels is not expected to be at the rate of much more than one level per year. Note however that levels are not linked to specific ages; year 1 is not necessarily full of children at level 1. In any given class of 7-8 year-olds, it will be possible to find children who are at level 2 or below in science, whilst others might be at level 4 or even level 5 in some aspects of science. The educational rationale for continuous assessment of children's level of attainment is that it helps teachers diagnose understanding and thus provide appropriate activities for all children, rather than work which makes a uniform demand and is thus inappropriate to a section of the class. In designing activities for children then, we should ask not 'what age are they?' but 'at what levels are they?'

This also implies that we are quite likely to find children at level 3 in the top end of KS1 (possibly in an infant or first school); in KS2 (in junior or middle schools); and in KS3 (at the lower end of secondary or high schools). Children at level 3 are required to show their attainment in relation to the same criteria. But does 'knowing' something mean the same thing for a bright 6-7 year-old as it does for an under-achieving 12 year-old? How do we interpret a criterion such as 'demonstrate knowledge and understanding of...' for children at different ages, in terms of what and how to teach it?

The Programmes of Study, which I have described as the 'templates' for planning science activities, are not themselves linked to specific levels. On the other hand, the Attainment Targets by which we have to assess children are. Thus if a child can 'sort materials into groups and describe them in everyday terms' we can say that the child is working towards level 2 in that AT. In order to know what level each child is at, therefore, our teaching will have to be organised in such a way that we can find out at some stage (and not necessarily at the same time for all children) if each child can group materials and describe them. This is the implication for organising teacher assessment. It is clear from the preceding paragraphs therefore that whilst the Programmes of Study are intended as the basis for planning science work, in reality the Attainment Targets and the practicalities of assessment are also going to exert a strong influence on what teachers teach. Some of these influences are discussed in the next chapter.

CHAPTER 3

THE IMPACT OF NATIONAL CURRICULUM REQUIREMENTS ON THE PLANNING OF SCIENCE ACTIVITIES

The main impact of the National Curriculum and its assessment may be fourfold, namely on the separation of process from content; teaching to the test; the difficulty of integrating science into topic work; and affecting the extent to which children engage in collaborative investigations.

As the example from electricity showed, whilst we are expected to teach in a way which integrates knowledge with investigation skills, the assessment process separates these in the way the statements of attainment are worded. It has been well-established for many years that most primary teachers are not very confident in teaching science; at the same time, it is also clear that most teachers want their children to do well, especially where public assessment is concerned. A danger, therefore, is that the level statements in the ATs will come to dominate teaching; and this could mean that we end up assessing the 'knowledge' attainment targets (ATs 2-4) separate from the investigation skills. More than that, in fact; because since *practical* skills are assessed in only one of the four ATs, (if at all, by the time this is published!) it is possible, even likely, that the emphasis on investigation will diminish, despite the intention that it should be given equal emphasis with knowledge. There is evidence from various countries that less confident teachers revert to more didactic, information-oriented methods in science when teaching topics in which their own subject matter knowledge is not well developed, and that they rely more heavily on textbooks in such circumstances (Carre, 1993). Thus the link between experiential, investigative science and concept development may be attenuated to such an extent that, for teachers as well as learners, it becomes lost altogether.

The changing nature of the Standard Assessment Tasks may also be reinforcing this trend. As already noted, these initially occupied a considerable amount of time and involved children even at KS1 in practical investigation. The need for teachers to provide such practical experiences in their day-to-day teaching was thus self-evident. However, successive Secretaries of State have reduced the investigative dimension at the expense of increased reliance on pencil-and-paper tests. The work of the Assessment and Performance Unit (APU) over the years, and similar bodies in other countries such as Kenya, has shown that such tests need not simply test recall of knowledge, and that they can for example eradicate bias in favour of urban, middle class children (Makau and Somerset, 1978). However, in other countries where such tests at the end of primary schooling are well-established, and in our own experience of primary leaving examinations such as the 11-plus (still present in a small but growing number of authorities), there is ample evidence to show that the nature of the test comes to dominate teaching, especially in the year immediately prior to the test itself. Moreover, tests increasingly take on a selection role, losing their function as diagnostic instruments serving teaching. Secondary schools, whether private, grant-maintained or state schools, are now in a position to select pupils according to their attainment in primary school, as the test case in Penrith in 1993 has shown; further selection is going to be encouraged as a matter of current government policy. Whether it happens or not, officially or unofficially, and on what scale, remains to be seen. Given the way education in England has always divided to some extent along class lines, the omens are not good.

We need now to look at how science, as prescribed by the National Curriculum, fits in to the primary curriculum as a whole. This issue has been at the heart of much discussion about primary schooling in recent years, well exemplified by the discussion in the 'Three Wise Men' report on curriculum organisation and classroom practice (Alexander Rose and Woodhead, 1992). In a section on 'subjects or topics?', the authors define topic work as 'a mode of curriculum organisation, frequently enquiry-based, which brings elements of different subjects together under a common theme' (p. 21). They point out that about 30% of work in primary schools is taught as single subjects, mainly music, PE, mathematics and some English. They also draw a distinction between Integration (bringing together subjects with distinct identities) and Non-Differentiation (not conceding that such distinctions are acceptable). From this, they assert that 'a National Curriculum conceived in terms of of distinct subjects makes it impossible to defend a non-differentiated curriculum'. Pupils must 'be able to grasp the particular principles and procedures of each subject and...progress from one level to another' (p. 22). They go on to claim that much topic work is and has been very undemanding in this sense, and debate the arguments for and against specialist subject teaching in science, without coming down in favour of one or the other. However, the same authors look favourably on the notion of the 'Semi-Specialist' teacher, who teaches his/her specialist subject but also has a generalist class teacher role as well as a consultancy role in relation to colleagues. Whatever roles teachers play, however, the authors of the report are in no doubt that improvements in classroom practice will only come about if the integrity of each subject is preserved and clear progression established within it.

Currently, however, the integrity of science is not always clear in all schools and classes, by a long way. An early attempt to look at how to make the National Curriculum work in primary schools showed powerfully that it was much easier to fit English and maths programmes of study into topics than it was to fit in the more specialised science content (ASE, 1989). And anyone familiar with primary school practice will be able to quote examples of how science knowledge and skills have often had to be twisted almost beyond recognition in order to fit a topic on The Tudors and Stuarts or Festivals. This kind of work is indeed a threat to the integrity of the subject. The outcome is often that the aspects which are omitted are precisely those which are most specific to science. It is usually not difficult, for example, to find opportunities in a topic for observation, measuring and recording skills, which are part of science but not specific to science. What is much more difficult is to find opportunities for children to engage in hypothesis testing, design and implementation of fair tests by controlling variables or replicating experiments to check on reliability of measurements; activities which are central to understanding what Bruner called the 'basic ideas' of science. Equally difficult, in a term spent on say Tudors and Stuarts, might be opportunities to incorporate valid work on environment, technology or electricity.

Obviously other opportunities for teaching these aspects of the science curriculum will arise, and study of the Elizabethan period may be the ideal opportunity to deal with various aspects of Earth and Space through learning about Drake's navigation around the world, and about push and pull forces in the same context. But – and I think it is a very big but – this leaves a great deal to chance, and may well deprive children of exactly that integral, cohesive view of science as one 'way of seeing the world' which has been regarded as essential. There may well be a topic structure which manages to incorporate the knowledge and skills of science in a cohesive way; and maybe that structure is a science structure! At the very least, we need to look at the schemes of work for each year in a school, and examine their content and methods in this light. But what we intend to teach and what actually happens in the classroom may often differ, and so often it is the harder-to-handle science-specific elements which are lost.

However, one effect of the increased attention paid to science in primary schools over the past 15-20 years has certainly been the increased amount of group practical work undertaken by children. It has become commonplace for 3-6 children to engage in such practical tasks often for a whole afternoon at a time; the advantages and drawbacks of such work in science have been the subject of a good deal of research, for example by Galton and Williamson (1992) and by Bennett and Dunne (1992). Clearly such work can foster discussion of the task and clarification of ideas, as well as the division of labour, modelling of manipulative skills and shared presentation of results, as discussed in greater detail in chapter 7. However, the success of such work has in some ways been its downfall, in the face of the demands of National Curriculum assessment. Early SEAC documents to help teachers with assessment acknowledged the difficulties inherent in trying to assess individual children's attainment whilst engaged in collaborative activity, whether by on-going teacher assessment or in SATs. The danger therefore is that this will lead to a gradual diminution of time spent on collaborative practical activity. And of course this diminution might result in concomitant disadvantages for bilingual children, children with other specific learning difficulties and others for whom the group practical activity is a particularly valuable learning context.

Making decisions about what to teach children in science

The above discussion about what science to teach, and how the National Curriculum imposes constraints on this, has hopefully given you a good deal to think about. The next step is to consider how to go about turning this into schemes and plans for the children you teach. At this stage, suggestions are provided as guidelines at a general level; how to turn these into more specific plans will emerge from chapters 4-6, which provide examples of science activities and ways for managing them in the classroom.

It is crucial first of all to **maintain the integrity of science as a discipline**. If in your school you do not teach science as a separate subject, this can perhaps best be achieved by organising work around science topics where possible. For example, the paragraph on electricity quoted on p. 33 could well serve as the spine for a term's work. It is not difficult to see how work in language, maths, history, geography, art, drama and music could relate quite naturally to practical activity on circuits: for instance, by investigating lighting and heating in the home and at school; role playing the dangers of fooling with mains electricity; meter readings and electricity bills; compasses, navigation and map reading; keyboards, loudspeakers, electric guitars; the introduction and social impact of electrical devices such as light-bulbs, etc. Many teachers have pointed out that it is much easier to find opportunities for language and maths work for example in a science topic, than it is to find opportunities for genuine science learning in a language or general topic. Thus the Science Programmes of Study are a good place to start in planning out programmes of work for your class or year group, to ensure that both the knowledge and investigation skills of science will be adequately dealt with, as well as those of other National Curriculum subjects which often come easier to teachers. One example of this, from a year 6 class, is provided in the following chapter.

It is then valuable to look at ways of interpreting the topic chosen by **focusing on 'real world' science issues** which children care about. For example, in Materials and their Properties, children have to investigate such properties 'including hardness, strength, flexibility and magnetic behaviour' and to 'relate these properties to everyday uses of the materials'. Where do children come across such properties? We might think about 'problems' they encounter, like the hardness of sweets (or ice-cream fresh from the freezer!), the stretchiness of tights or the requirements of an effective cycle helmet. They can use chromatography to do forensic tests (which pen was used to sign the forged cheque?!); or do consumer tests to find the bounciest ball, the socks that get holes quickest, the sudsiest washing-up liquid, the stability of a baby's high chair, or the best (stickiest?) glue. This is real-world, applied science, and helps children to realise that they themselves can test ideas by using the evidence of their senses. Hopefully, it also reduces gullibility and dependence on 'experts' in white coats, and is therefore empowering.

Thirdly, it is important, as well as illuminating, to plan to start each topic by **finding out** (another word for assessing) **what children already know and believe**, so that you can build on and develop these ideas. Your early activities should focus deliberately on this. A good example (Crossley, 1991) is of a teacher beginning an activity by allowing children to place toys at the top of a slope, and then asking them 'why do they go down the slope?' The answers (from 'gravity' to 'because it has wheels') gave a clear indication of the range of understanding, which was helpful in choosing the subsequent activities for children. For example, a child who thinks only toys with wheels can go down slopes may need to investigate things which slide; whilst the child who already has ideas about gravity may need to predict and then investigate what happens if you make the slope steeper. This is a clear indication that more than one kind of activity is likely to be needed at a given time, to match the needs of the class, which in turn has important implications for planning. For this diagnosis at the beginning of a topic, you might equally well start with open-ended questions such as 'tell me some of the things that might affect...'; 'what is likely to happen if...'; or 'in what way might you...'. Once you convince them that you value their ideas, you will not be short of suggestions.

In the resource book on Science Explorations (NCC 1991) a useful section on 'revising science activities' suggested examples of how to turn closed activities into investigations which help children explore skills and concepts. For example, a question such as 'will these gloves keep your hands warm?' might evoke guessing and conjecture, but little in the way of suggestions for a controlled investigation. A more productive way to express the problem might, they suggest, be; 'How can we find out which gloves are best at keeping your hands warm?' In other words, questions are being turned into 'testable' questions.

Beliefs are as important as knowledge, and often cannot be separated. For example, if children think that electricity 'is hot', or 'comes from the shop' or is 'made by God', are these misunderstandings, errors of fact, or beliefs to be respected and explored? Working with batteries and bulbs and encouraging them to talk to each other about what happens may serve to develop their ideas in a more profitable direction: the intervention of you the teacher as listener, questioner, provider of words for evolving concepts, is very important. At the planning stage in a science topic, some likely stumbling blocks can be anticipated by studying research findings such as those of the SPACE project exemplified above. However, there is no substitute in the end for listening to your own children; and not in a hapazard way, but by deliberately eliciting and exploring their ideas and accepting them without being dismissive. Some of the ideas now commonplace in science were thought weird, when first proposed, by the general public. Do you remember the first hovercraft? In a recent issue of the 'Independent' (12.4.96) newspaper is an article reporting that 'Scientists' have isolated a gene which prolongs life; it may soon be possible to extend 'normal' human life to around 150 years. Here is a scientific issue which all children will want to debate. Is it a good idea? They may well have watched 'Gulliver's Travels' on TV and been provoked by Swift's comment on the implications of having eternal life; they will have ideas which it is important to explore. What would it mean for the population of the planet, retirement age, food production, birth control?

The point is about children's confidence and self-esteem; about not destroying their willingness to hypothesise and play with ideas. For the same reason, therefore, it is important in planning science work to **encourage the idea that we are all scientists**; that science is a normal, fun activity, and not (as in the media stereotype) for 'mad Scientists' only. We can do this by doing real science, such as the consumer tests described above; and by simply being scientific about the day-to-day occurrences around us, rather than simply accepting what 'experts' tell us. Questioning adverts is one way to do this. (Does Weetabix really give you all that much more energy? A survey of the charts on cereal packets will soon sort that out).

They can also study and use familiar, natural phenomena in a scientific way. (Can you design and make a working sundial? How accurate is it? How can you tell? Would it be more accurate if you made it twice the size?) Or to make life easier. (How can you use your science knowledge to cut down the number of complaints about the noise of the ghetto-blaster in your bedroom, or to make it stay dark longer for my 'sleep-in' on summer sunday mornings? How can I help gran open tight jampot lids?) Then there are things which just interest them (What difference do colour filters actually make to photographs? Did we really have more rain this June than last? Does Max Factor lipstick really stay on longer?) All these questions relate directly to sentences in the programmes of study for KS2. They can all be investigated scientifically.

Some other questions cannot, of course; and it is very important for teachers and children to be able to recognise which are which. All children want to know about where we came from, how the universe began, what happens when we die; is the earth really getting warmer and are the ice-caps melting, is the ocean rising, how can we stop it; how can we prevent oil-tankers polluting beaches and killing wildlife;

will we catch CJD if we eat burgers?. It would not be wise to give children the idea that science can easily or quickly 'solve' any such problem. But it would be just as sad if they learned that 'they' can't tackle any real science problems. Womack (1988) has reminded us that there are still many problems left which children are interested in and can tackle easily. Children become scientific by doing science, by learning to talk the language of science together; not by being told to accept what someone else tells them is true, in a 'foreign' language.

One implication of all the above suggestions, then, is that **you need to understand the science implications and potential of the Programmes of Study**. This means a clear grasp of the concepts and principles; working knowledge of the science skills demanded; and the ability to turn untestable questions ('why can you hear a lot of things at once?') into testable ones ('how might we tell where the source of a sound is?'). A helpful way to tackle this might be for you to produce your own 'concept map' when planning to turn a section of the programme of study into a plan of work. The recent channel 4 series 'Making Sense of Science' has useful concept maps of each content area in the printed back-up material

An example of which is illustrated below.

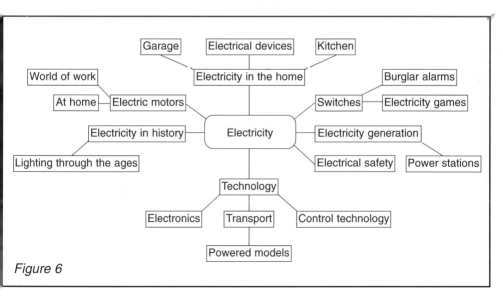

Figure 6

What this does is remind you of the ideas to be learned and the relationships between them; the **propositions** which children need to generate to show their understanding. So you yourself must be able to generate your own ideas and propositions, as a 'self test' of your understanding, before you plan a scheme or programme of work.

All teachers are aware of the requirement for continuous assessment of children's science attainment, but this does no mean that all teachers accomplish it effectively. Our own evaluation of the early years of assessment at KS1 (Desforges, Mitchell and Peacock, 1992) suggested that some teachers took it in their stride whilst others made very heavy weather of the process, resulting in stress and exhaustion. So in turning a programme of study into a scheme, feasible opportunities for assessment have to be built in right from the beginning.

Many new published schemes offer considerable help with this. For example, in the Bath Science materials, each section has assessment boxes such as the following (see Figure 7).

SEAC and subsequently SCAA have also produced a variety of publications from the general to the very specific, such as the 'Assessment Matters' series (EMU/SEAC), the School Assessment Folders (SEAC, 1992) and the SCAA Exemplification of Standards for core subjects (1995) which illustrate children's work at different levels in science. All of these allow you to compare your own approach and standards with those of teachers elsewhere.

The intention is to incorporate authentic activities which provide assessment information without disrupting learning and overburdening you, the teacher. In other words, assess them in the way you want them to learn.

Planning science work is also dependent on **resources** available. By this I do not mean bunsen burners, test tubes and spatulas, although these might conceivably be useful on rare occasions (if you have piped or bottled gas in your classroom!). By resources, I am thinking more of a bank of good ideas which work; raw materials to work on; people with skills, interest and enthusiasm, 'infectious know-how'. The idea that doing science is dependent on having a laboratory with specialist equipment is just as dangerous a stereotype as that of the white-coated boffin. People who are a whizz with hot air balloons, nature trails,

Figure 7

ASSESSMENT

If the children can demonstrate that they can link common materials to their properties they have achieved Sc 3/3a. If they can also explain how a manufactured material, such as nylon, is made they can also achieve Sc 3/4b. For AT1, the children can demonstrate the following levels in each strand:

Level	Strand (i) planning	Strand (ii) experimenting	Strand (iii) evaluating
2	suggests that anoraks you warmer than blazers because they are of thicker material	tries on coats of different thickness of material and describes which keeps them warmest	explains that a particular coat keeps them warmer than others
3	suggests that one material is better at keeping an object warm (or dry) than others and goes on to test this	uses a container of hot water and measures its temperature against time	consistently uses the same starting temperature and degree of covering of the container, states that material A kept the water warmer than B because it stopped the heat escaping to the air
4	suggests that anoraks are an insulating material and predicts that thicker layers of this material are better than thin layers	uses appropriate instruments and identifies temperature as the variable to be measured, the material thickness as the variable to be changed and identifies what other variables should be controlled	explains that thicker materials are more effective insulators
5	considers two or more factors such as thickness of material, tight, or loose fitting, single or composite material and ensures that the starting temperature of the beaker is around body temperature of 37°C	organises a fair test so that the temperature of the surroundings reflects winter temperature (0-10°C) and suggests which variables should be controlled (e.g. container shape and size, volume of water, area of material used)	explains results in terms of properties of the fibres that make the material or ability of the material to trap air as an insulating layer.

bread making, Australian stick-insects or constructing buggies are much more useful. So are source books with tried and tested ideas which draw on wider cultural and global perspectives. Ideas for investigating utensils (what determines the best shape for a cooking pot or a rolling pin?), keeping cool (as well as keeping warm), and other peoples' weather, fruits and fuels as well as ours, are more valuable and interesting than culture-bound work on a rubber band or a cup of tea. Resources for such work are increasing all the time, and can be found in the catalogues of various organisations such as Development Education Centres and OXFAM (Peacock, 1991). So the word from the streets is: accumulate resources, develop your network of contacts, build your ideas bank. And don't forget books, ideas and materials from other countries and cultures, in doing so. I was recently given a wonderful picture-book of children's stories from Hong Kong called 'Weighing an Elephant' (He Youzhi, 1981). There is a whole term of fascinating science in it! (How *would* you weigh an elephant?! Could you use a boat and a pile of rocks....?)

Lastly, in turning a curriculum prescription into a scheme of work, it may be invaluable to **consult and inform parents at the planning stage**. Parents can be a huge resource, in terms of support with science work at home, raw materials and background knowledge. Studies by teachers have shown that parents often 'do science' with children at home without realising it (Parker, 1990). You might find that hot-air balloonist you need! However, many parents are still 'backward at coming forward' with help, especially in science where many still have very hazy ideas about what is involved, as discussed in chapter 7. So tell them what you're going to do, and encourage them to co-operate. The spin-off from this may be more interest amongst children, more time for you to assess when a parent helps in the classroom, new ideas for 'real' activities, more materials to work with, and crucially, more help from parents for children in as well as out of school.

It might also mean spreading the idea that science really is 'for all'; that it is what we all as humans can do, an OK way of dealing with the world around us. Values and ethics are on the agenda, not only when we deal with animals in the classroom and need to care for them and return them to their natural state, but also when we look hard at our energy and water consumption (how much water do you run when you clean your teeth?). A scientific study of the kind of work mums (and dads) do at home might be a good way in to Forces and Energy. So often in the past, scientists and teachers of science have taken the subject out of this human, ethical domain; but the children I know, all know, probably without realising it, that you can't separate the two. Let's put it back, for their sake, and possibly for the sake of the future of the planet.

CHAPTER 4

THE REALITY OF SCIENCE IN A PRIMARY CLASSROOM

This chapter focuses in detail on the work done in science by two parallel year 6 classes in an urban primary school in the summer term, 1993. Description will stress how the work fitted with the topic for the term and the National Curriculum; the planning and execution of the science activities, including constraints and choices made; management of 'classroom' work, and evaluation of the outcomes in terms of what the children did, learned and felt. This description of actuality will then be the basis of the discussion of lessons to be learned in chapter 5.

Planning the science work for a term

Figure 8 sets out the term forecast for the work of year 6, on the topic 'Britain since 1930'. Such a forecast is required by the Headteacher, and was produced by the two class teachers, Nick and Linda, in consultation, at the beginning of the year. Three areas had been chosen for attention in science, namely Smoking; Food Chains/Ecology/ Habitats; Electricity. Linked to these would be some of the work in Technology (Household Technology, e.g. electrical implements).

The science dimension was to be taught mainly by Nick (who happened to be the science co-ordinator for the school) to each class separately, on half a day each week. Thus on Wednesdays, one class

Figure 8

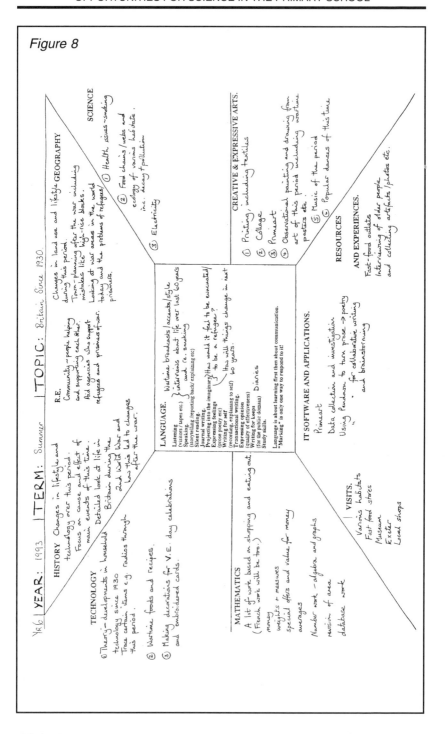

Figure 9

TOPIC Britain Since 1930's   **TERM** Summer 93   **SUBJECT** Science.

AT	Content/Activity/Skill	Resources	Outcome/Evaluation
AT2:	The effects of cigarettes and drugs/alcohol. The use of cigs. in wartime compared to now. Asking friends and relations if they smoke or not and surveying drinking smoking habits.	Cigarette Test Kit. Literature about alcohol/ tobacco from doctor.	
AT2:	Competition between living things within a habitat and the various needs of animals and plants. How might pollution affect a food chain at one particular level and hence at levels above and below? How the process of decay affects the life of living things. How decay takes place.	Visits to Stoke Woods + Dawlish Warren. + Topsham Bowling Green Marsh.	

would have science from 9.30 until noon; the other class would then have half an hour before lunch and the whole afternoon for science.

Prior to the Easter holiday, the science component of the year's work was amplified by Nick as shown in fig 9 above. Note that at this stage, the implications for resourcing and visits are being developed, and links to the National Curriculum attainment targets are specified. It was also agreed between the two teachers that the first half term would concentrate on the Smoking and Habitat work, whilst the second half term would be given over entirely to work on Electricity.

The school has its own minibus, and it was intended to use this for visits to different habitats during the first half of the term. In terms of numbers of children, this meant taking out one third of the year-group on each of three successive Wednesdays, with one teacher and a helper on each occasion. This constraint was a key factor in determining the science programme for the first half-term, as school-based activities had to be planned for two-thirds of the year-group in relation to the date of their field trip. Hence the first outline plan of activities looked like this:

Figure 10

WEEK	GROUP 1	GROUP 2	GROUP 3
WEEK 1	PREPARATION FOR FIELD TRIP	PREPARATION FOR FIELD TRIP	PREPARATION FOR FIELD TRIP
WEEK 2	FIELD TRIP	SMOKING ACTIVITIES	SMOKING ACTIVITIES
WEEK 3	FOLLOW-UP TO FIELD TRIP	FIELD TRIP	FURTHER WORK ON SMOKING
WEEK 4	SMOKING ACTIVITIES	FOLLOW-UP TO FIELD TRIP	FIELD TRIP
WEEK 5	SIMULATION; PUBLIC ENQUIRY ON SMOKING	SIMULATION; PUBLIC ENQUIRY ON SMOKING	SIMULATION; PUBLIC ENQUIRY ON SMOKING

* in weeks 2, 3 and 4, the whole day was given over to the activities described, whereas in weeks 1 and 5, only half the day was allocated to science in each of the two classes.

It is apparent from this timetable that the three groups would have different experiences, even without allowing for the vagaries of the weather. However, the differences in practice would not turn out to be as great as the table indicates. For example, group 3 did not miss out on follow-up work to their field-trip, as the table implies; this was undertaken on other days, between the fourth and fifth Wednesdays. Similarly, the extra 'Smoking' work covered by group 3 was also done by groups 1 and 2 on these 'in-between' days. What the table does show clearly, however, is how one resource factor (transport, in this case) can have a considerable influence on the planning of work in a topic, and on children's experiences.

In the second half of the term, the children would revert to being taught in their two normal classes for electricity work. The plan for this half of the term was more straightforward, but had three clear stages, namely:

Week 7: a period of revision and assessment of previous work on circuits and household electrical implements;

Weeks 8-11: new investigations to extend their learning of concepts and skills;

Week 12: display and consolidation of things produced and ideas learned.

Decisions about exactly what to do on each of these five Wednesdays was not made until almost halfway through the summer term itself. Work for the first half term was however mainly planned before the Easter holiday, and finalised during the last week of the holiday by the teachers involved.

The rationale for the activities chosen

Some of the initial criteria for choice were as follows:

- children must have opportunities to undertake investigative work and, where possible, play a part in the design of the investigations, so that these would seem relevant and interesting to them. (Some authors might talk here about 'ownership' of activities, but in this case it implies rather more than the teachers were hoping for)

- they must be responsible for making first-hand observations and for finding ways to record these more systematically than in the past

- work in the classroom should encourage children to interpret (i.e. make sense of) their observations, and to report what they found to others by different means of communication. An important feature of this would also be to develop listening skills and confidence

- children should acquire ideas through a variety of media, such as first-hand field observation; reading and research in reference books; watching video material; participating in collaborative group tasks and discussion; creating and making things; communicating ideas

- specific concepts needed to be understood by the children, such as Habitats; Food Chains and Food Webs; Identification Keys; Circuits; Conductors; Resistance; Symbolic Representation

- management of the activities should at all times seek consciously to discourage gender stereotyping.

Week 1: Preparation

In order to provide children with the opportunity to take part in designing activities related to the study of the habitats, Week 1 was devoted to a 'Dummy Run' in the school grounds. Children were put into groups and asked to do one of the following activities:

- mapping an area of the grounds
- making and using a quadrat
- using a plant identification key
- using a bird identification key
- doing a ground wetness test.

In each case, the activity had been designed by the teacher in a way appropriate to the school grounds. An example of the materials provided for children to do these activities is shown in figure 11.

Figure 11

You will need:

An outline sketch of the area
Clipboard
pencil
spare paper for notes

What you have to do

Decide what important features to put onto your map. (this may include types of vegetation; areas of water; walls; paths, etc.

Draw these features where they go on the map. Keep it simple; don't put too much on, just the important things.

Make some notes which explain why you think the living things are where you saw them. For example, why the birds are in the trees, or by the water.

Compare your results with others. Look at how they are similar and different. What did you learn from doing this?

Having tried an activity, each group was then asked to re-design it for use in a woodland or marshland environment. This required them to consult appropriate reference books (woodland birds, marsh plants, maps). The group would then use their modified activity on the following week, in the field. Figs. 12 to 14 are examples of some of the activity sheets designed by the children.

Copies of the group-designed activities were made, so that all children in the group would have a copy for the following week.

Children worked in groups of four. The parties for the three field-trips were chosen by the teachers with ease of management in mind: this involved taking into account abilities, friendship groups, children who

Figure 12

Ground Wetness Test

*FIRST OF ALL YOU NEED -- 5 PAPER TOWELS,
ELECTRONIC SCALES A PENCIL AND A SHEET
OF PAPER. YOU CAN RECORD YOU RESULTS ON
A GRID LIKE THIS --*

Place	Weight in grams
e.g. Pond	7 grams

*CHOOSE 5 PLACES WHICH YOU THINK THE
MOISTURE OF THE GROUND WILL BE
DIFFERENT.
AT EACH PLACE YOU WILL NEED TO CHANGE
THE PAPER TOWEL WITH A CLEAN ONE. PUT
IT ON THE GROUND AND PRESS ON IT FOR 5
SECONDS. THEN THE PAPER TOWEL WOULD
SOAK UP THE MOISTURE FROM THE GROUND.
THEN YOU PUT THE PAPER TOWEL ON THE
ELECTRONIC SCALES. THE NUMBER THAT
APPEARS IS YOUR ANSWER.*

BY Aaron .H.
Andrew .C.

tended to have difficulties in working under limited supervision. However, within these larger groups, children were free to choose their working groups. At the end of the preparation day in week 1, the teachers met to decide who would do what on week 2, such as who would be responsible for which materials, and how the school-based 'Smoking' activities would be organised.

Weeks 2/3/4: the Field Trip

On each week, the group of 16 children and 2 adults left school at 9.30 a.m., and spent the morning in a Forestry Commission woodland habitat, about 3 miles from the school. A picnic lunch was eaten in the wood at mid-day, after which the group moved to a marshland environment close to an estuary, a journey of about 6 miles from the wood. There was time for about 2 hours work at each habitat. On week 3, unfortunately, the rain was so severe in the afternoon that the marshland visit had to be cut short after only 30 minutes of getting soaked! This meant that the follow-up programme for one group had to differ from that of the other two groups.

In the morning, the children assembled in a clearing in the wood and were given reminders about 'dos and don'ts': staying within earshot, staying as a group, not shouting to disturb animals, not damaging plants. Some suggestions were given about where best to attempt the different activities. Groups were then left to try out the activities they had designed the previous week. Some inevitably found that their activities as designed needed modifying, and went on to do so. Others had difficulty with the task (for example, the mappers often chose too large an area, or an area in which it was difficult to move about or to see features) and in these cases, the adults assisted the groups in solving their problems. There was inevitably off-task activity – freedom in a woodland environment inevitably provides opportunities for all kinds of non-scientific fun! – and staff had to find an appropriate balance between allowing children to enjoy and explore the environment whilst at the same time maintaining a focus on the chosen activities. This raised an important point for discussion between staff namely, the way in which the National Curriculum programmes of study have influenced the purposes and specific objectives which

Figure 13

How to use the bird-key

First you look around and when you have spotted a bird you decide whether it is pigeon size, black bird size or sparrow size. Having found the right sheet find the start here point. Now look at the boxes joining onto it. Something will be writen in each one. Decide which one matches your bird best. you keep doing this until you come to the name of the bird. There will be no boxes joining this box apart from the box that you have come from. Now you have discovered the name of the bird!

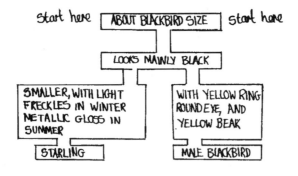

BIRD KEY
BLACKBIRD-SIZED BIRDS

start here | ABOUT BLACKBIRD SIZE | start here

LOOKS MAINLY BLACK

SMALLER, WITH LIGHT FRECKLES IN WINTER METALLIC GLOSS IN SUMMER

WITH YELLOW RING ROUND EYE, AND YELLOW BEAK

STARLING

MALE BLACKBIRD

Figure 13 continued

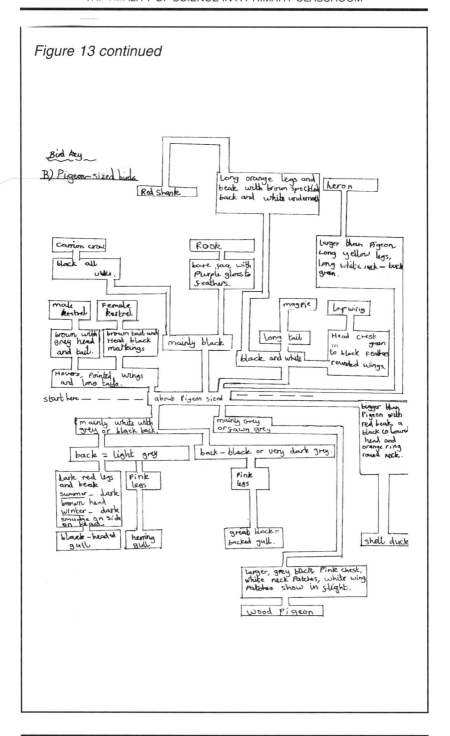

Figure 14

Instrutions For using the Quadrat.

Equipment: A Quadrat
A clip board
A pen or pencil
A Woodland and marshland chart

What to do: First get a quadrat and throw it on the ground (make sure when you throw it down it makes a square) then see what plants and grasses are inside it. Get a book on plants and trees and try to match the leaves or plants to the ones in the book. Then when you have found out every plant or leaf that is inside the quadrat write all the things you found in your quadrat on to the appropiate chart according to where you are. On your first throw write what you see the first time in the first section On your second throw write what you see the second time in the second section On your third throw write what you see the third time in the third section. You will need to do this 3 times in both woodland and marsh. Now you are set to go.

teachers have when taking children into natural environments such as this, and the consequent implications for intervention.

In the open marshland habitat, the focus of work was similar to the morning, but the teacher's role was different. A main purpose of the day was to carry out similar activities in two habitats in order to compare results; hence groups repeated the ground wetness test, mapping, use of quadrats and identification keys. Supervision of activities was easier, since all children were visible all the time; but the distraction of concealment in the wood was replaced by the temptations of open water in the marsh. Often, children who had not been fully engaged with the task in the morning were engrossed in the afternoon, and vice versa, sometimes because one or other of the habitats was more familiar. Some, for example, had been taken birdwatching on the marsh by fathers or older brothers, and could therefore use their existing knowledge and skills.

The role of the teacher in this case was not simply to keep children on task and to draw attention to things they might miss, but to raise key questions which would help children make comparisons: such as 'did you see... this morning?'; 'is it different/larger smaller than...'; 'were there more/less of these this morning?'; 'what do you notice about... (the ground/plants/ creatures etc.) that you didn't notice this morning?' It was also important to ensure that children kept records which would be useful to them during follow-up work in school, both in relation to their particular activity and also to interesting observations in general.

Of course, doing this effectively with several groups scattered over a marsh involves compromises. At any time, some groups will be without teacher input, and need therefore to know quite clearly what your expectations are, in terms of tasks to be completed within a certain time, and in terms of safety. It also means that the teacher has to 'think on her feet' concerning unplanned activity; for example, studying a heron fishing, trying to locate skylarks' nesting areas. Should children be allowed to pursue these or not? And in the case of the very wet afternoon, how soon to curtail the activities and return to school, and what to do on returning. It is not a question, here, of knowing the 'right' solution in advance: often there isn't one. More important from the point of view of the inexperienced teacher is to evaluate the consequences of your choice, and learn lessons which will guide decisions next time. In the case of the wet afternoon, for example, children returned to school with about 45 minutes of working time remaining, and spent this making finished copies of their (rain-spoiled) records, in preparation for follow-up work the following week. For most children, this proved to be worthwhile and satisfying, as they had many bits of information to assimilate from the morning's work. For the teacher, it meant taking account of the fact that follow-up would need to be managed differently for different groups, and some note of this needed to be made and communicated to the other teachers. Hence a weekly meeting was held to share this kind of information, and to plan the detail of the coming week's science work, on Mondays after school.

Weeks 3 and 4: Follow-up activity
There are various appropriate functions for follow-up work. First, some things simply have to be completed; drawings, tables, descriptions of observations. Second, children need often to make 'neat' versions of their field records, to provide them with the satisfaction of a product they are pleased with, and sometimes to enable these to be displayed. Third, children need to reinforce new words, concepts, relationships; to use these in writing and discussion and new contexts, to help with restructuring of ideas and provide confidence in using terminology. And fourth, some children need to consolidate understanding through discussion and drawing out of inferences, whilst others need to have the opportunity to extend their thinking, pursue questions they have raised, to 'be stretched' in everyday language.

The teacher's role in the first two of these can be fairly undemanding; perhaps providing appropriate materials and keeping them focused on the task, reminding them of standards expected and deadlines to be met. However, the teacher's role in the third and fourth functions is crucial, and it is this monitoring, mediating and challenging function which inexperienced teachers often find most difficult. Essentially, it means identifying where children need 'scaffolding' so that their learning can move forward, and providing the appropriate input, for example through the medium of a structured conversation. Such techniques of mediation will be discussed in chapter 5.

Weeks 2/3/4/5; work on Smoking

As figure 16 indicates, this section related to earlier work on health education and in particular to the growth in cigarette sales since the thirties. Much of the knowledge content derived from up-to-date statistics from the local branch of the Association for Smoking and Health (ASH); from the 1993 survey by the Royal College of Physicians (Smoking and the Young; Children's Right to be Free from Tobacco Charter), from the County Health Education Adviser and from the Health Education Association Research Unit, which also provided local statistics on the growth of smoking in children. The purpose of using first-hand data in this way was to enable children to interpret and make sense of the figures, to discuss the implications and begin to form an opinion of their own on the health implications.

This first-hand research was made more immediate by use of video material and practical activity to illustrate what happens during smoking. One way of observing this in practice was used, as illustrated below.

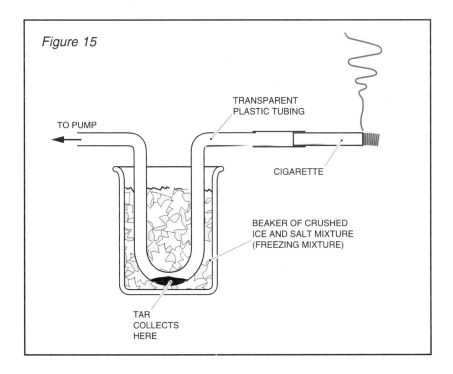

Figure 15

TO PUMP

TRANSPARENT
PLASTIC TUBING

CIGARETTE

BEAKER OF CRUSHED
ICE AND SALT MIXTURE
(FREEZING MIXTURE)

TAR
COLLECTS
HERE

Figure 16: A CHARTER
CHILDREN'S RIGHT TO BE FREE FROM TOBACCO
Royal College of Physicians 1993

Survey 'Smoking and the Young' published 1992 by the Royal College. Increased health risks due to smoking;

Smoking and Pregnancy
Effects are;
– infertility
– miscarriage
– low birth rate
– preterm birth
– perinatal mortality

1993; 1 in 3 pregnant women smoke, 25% of these remain as smokers at birth. Difficulties for health visitors in giving advice, easy to give judgemental advice which alienates people and they do not come to antenatal sessions.

Passive smoking in Early Childhood
Effects are;
– cot death
– pneumonia
– asthma
– glue ear
– hospital admissions
– school absenteeism.

Children and Smoking
450 children take up smoking each day. 1 in 4 school leavers are smokers.

Effects are;
– respiratory infections
– asthma
– time off school
– cancer risk
– time off school

Why do they do it?
– initiation
– experimentation
– regular smoking
– addiction

Main reasons for initiation and experimentation are risk-taking, rebelliousness and poor self-image. Up to 12/13 children well aware of health risks but 14 onwards dramatic increase in uptake of smoking. Why?

Major factor could be social reinforcement, approval of adults, peers and siblings has a strong influence. Do parents teach children how to smoke and set attitudes? Health hazards seem to have a low place in decision making process.

Environmental Influences on Smoking Behaviour
– non smoking teachers and schools with no smoking policies
– smoking in public places
– availability of cigarettes
– price
– tobacco advertising.

Gives rise to questions about best ways of targeting to bring about reduction in smoking uptake amongst children. Which would be the best influence to target? Children's and parent's views on each of these influences would produce some interesting data.

1993 Charter

Children have the right to ...

– be free from the effects of tobacco when in their mothers womb.

– be brought up in a home that is smoke free.

– expect that doctors, teachers and all those caring for them will set a good example by not smoking.

– schools, youth clubs and public places that are smoke free.

– be taught about the impact of smoking on health and well-being.

– be taught how to recognise and resist pressures to smoke.

– not be sold cigarettes and other tobacco products.

– be helped to remain non-smokers by the high cost of cigarettes.

– be free from any form of tobacco advertising and promotion.

– live in a community where non-smoking is the normal way of life for all age groups.

and to expect that public policy will reflect these rights.

(*Royal College of Physicians*)

In week 2, all children not on the field trip undertook the same work, i.e. they watched the video, discussed it as a class, and worked on the analysis of statistics in groups.

During weeks 3, 4 and 5, all children also undertook a practical activity with the 'smoking machines', to demonstrate the nature of the bi-products from burning tobacco. This work differed from group to group, some spending more time on it than others, some using both machines, others only one. This was an inevitable compromise which had to be made due to the pattern of the field visits. However, use of other times in the week kept the difference of experiences to a minimum.

As with the work on habitats, it was felt that the children's new ideas and developing views had to shared and consolidated in some way, and this was done in week 5 by organising a 'Public Enquiry' into smoking. This role-play simulation was done by each class separately, and took half a day each. A judge and jury cross-examined ten witnesses (see Fig. 17 for full list) each of which had a 'brief' which helped them devise and present their case. All the class took on one or other of the roles: 30 minutes were allotted to preparation, and an hour to the enquiry proper. The children took this remarkably seriously, entering fully into the role play, and saw it as the highlight of the work on smoking.

No attempt was made to 'write up' this experience; the intense debate and the development of listening skills were themselves the intended outcomes, as well as the 'rounding off' of this section of work, before moving on to new work on electricity. Fortunately, the enquiry concluded by a majority that smoking should be banned!

Weeks 7-12; work on Electricity

During the second half-term, all the science work was devoted to this topic. Without field-trips, it was possible to teach each class for half a day; however, to equalise the time given to each class, the morning session was completed at 12.00, allowing the afternoon group half an hour before lunch plus the whole afternoon.

Figure 17: List of roles in Smoking 'Public Enquiry'

Joe Grower, the Producer
Dominic Smart, the Advertising Executive
Harvey Woodbine, the Manufacturer
Wendy Packer, the Worker
Vanessa Cash, the Tax Inspector
Liz and Peter Proud, the Parents
Rachel and Daniel, the Proud Kids
Tracey Drag, the Smoker
Harry Floggit, the Newsagent
Joanne Cure, the Doctor

THE WORKER

My name is Wendy Packer, from Bristol. I've worked for old Harvey Woodbine for 12 years now, packing cigarettes into cartons of 200 for the duty free shops, then putting them into big cardboard boxes. It's a boring job, sitting all day next to massive pile of fags going past on a conveyor belt, stuffing handfuls into packets. I've got used to the smell, but my husband Melvyn says I stink of tobacco when I get home.

The job isn't brilliantly paid, but we're OK, and I've had no reason to complain about the bosses. Though they laid off a few hundred last year, and you never know if it will be your turn next. We get 200 free fags with our pay packet every week, which suits Melvyn fine, because he's a heavy smoker. I've tried to get him to cut down, but he says it's a pity to waste them. He had to see the doctor last winter because of his asthma and his lungs; I'm getting a bit worried. Still, there's not much other work for me round here, and we need the money.

As indicated above, the rationale for the term was worked out before Easter, but final decisions on activities were not made until after completion of the Smoking and Habitat work. The precise activities chosen were as follows:

- Revision work on constructing circuits using electricity kits incorporating cells, bulbs, bulb holders, switches, buzzers, wire and crocodile clips

- Assessment and development of children's ability to draw diagrams of the circuits which they had constructed

- Open-ended construction activities such as making a switch and bulb holder which worked in a circuit

- Using all the above skills to construct and describe a working model of Traffic Lights.

The precise timing of each was not determined in advance; for example, the work on drawing circuit diagrams was incorporated into each week's practical activity, children gradually refining their techniques and use of symbols.

Week 7; Diagnosis

The first week of this topic was deliberately diagnostic. Electricity kits were provided for each group of 4 children; questions were asked about terminology ('what do you call this?') and procedures ('what must you do in order to...?'). A video intended to teach these concepts was used, without the sound commentary, as a vehicle for asking questions to 'test' children's knowledge and understanding. Domestic appliances were also brought in and (in the case of the toaster and torch) dismantled to show how they were connected, and where the electricity went.

Groups were then asked simply to make the bulb light and the buzzer buzz, and to draw a diagram of how they achieved this. There was inevitably a wide range of previous experience of electrical circuits, ranging from two boys who were well advanced in home electronics to others who had fears about electric shocks. Teachers' observations revealed that;

- several groups had problems because of poor connections, and did not easily diagnose this; usually, the explanation given was that the bulb or buzzer 'didn't work'
- the fact that buzzers were unidirectional was also confusing to some children, who did not think to try them the other way round
- the words 'battery' and 'cell' were used interchangeably by children
- connections were sometimes made to the wrong place because many wires were being used and were allowed to cross over each other
- some children thought that the red and black wires served different purposes, and could only be connected in certain places
- children's drawings of their circuits were 'life drawings' which attempted to make bulbs and cells look realistic, whilst obscuring the important details of how and where they were connected. Symbols were not used in these diagrams. (see examples in figure 18).

Making Switches and Bulb Holders

These diagnostic observations were taken into account in deciding how to present the activities in week 8. Having used the switches and bulb holders from the kit in week 7, children were asked to make a switch or a bulb holder of their own and to incorporate it into a circuit. Several groups decided to divide so that two children made the bulb holder and two the switch. Materials provided included card, scrap wood, corriflute, paper clips, butterfly clips, foil, drawing pins, plasticene.

Children found the switch easier to make than the bulb holder, because of the difficulty in making good connections with the latter. This meant that all groups managed a switch, but not all succeeded in making a bulb holder which worked. A common switch design used card, butterfly clips and a paper clip as illustrated in figure 19.

Figure 18

Wednesday 9th June 1993.

Electricity Problems

How to light a light bulb and being able to switch it off and on.

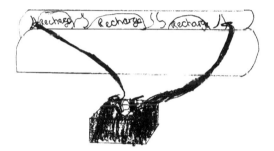

Is the red and black wire were swaped around the light would light up exactly the same way.

Figure 19

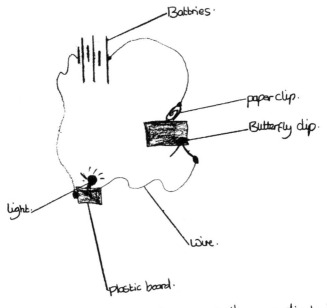

Toyah, Katie and Kelly.   Wednesday, 23rd June.

Electricity design.

To make this work you have to connect the paper clip to the butterfly clip.

As an extension activity, when groups had completed their switch, they were asked to adapt it as a two-way or three-way switch, so that one or other of two or three bulbs could be switched on at a time. This activity was introduced with the forthcoming idea of traffic lights in mind, although this was not mentioned to the children at the time. Most children quickly grasped that adding an extra butterfly clip or two could have the desired result; some however continued to struggle with problems (such as trying to use plasticine to hold wires together) and were resistant to alternative ideas.

At such a stage, a crucial decision for the teacher is whether to intervene to prevent frustration and failure, or whether to allow the children time to convince themselves that there is a better way to do it. The decision needs to be informed by many things; time available, group dynamics, knowledge of individual abilities, whether or not the finished product will need to be used later.

Based on the drawings from week 7, children were again asked to draw their circuit, but this time symbols were introduced. Children generally accepted the use of symbols for cells, bulbs etc. easily, but insisted on drawing the switch because it was, in effect, 'their' creation, and not a 'real' switch. They also tended still to draw wires as they observed them on the table, i.e. often twisted and overlapping, instead of as straight lines for simplicity. Again this was noted to be dealt with later.

By this stage, children were relatively confident with circuits and switches and the need for good connections. Where children finished work quickly, other extension activities were introduced; for example they were asked to incorporate other materials into their circuits, to see which did and did not conduct electricity, such as the 'lead' of a pencil.

This also helped to introduce the idea of Resistance, important because several children had experienced short-circuits and hot contact points, even tiny shocks, through the way in which bulbs had been connected, and the idea of the path of least resistance had been discussed. It was also important, at the time children experienced the physical effects of short-circuits, to discuss the safety issues and help them overcome any fears associated with electric shocks, fires and such things, as this can be one of the major barriers to children's confidence in learning about

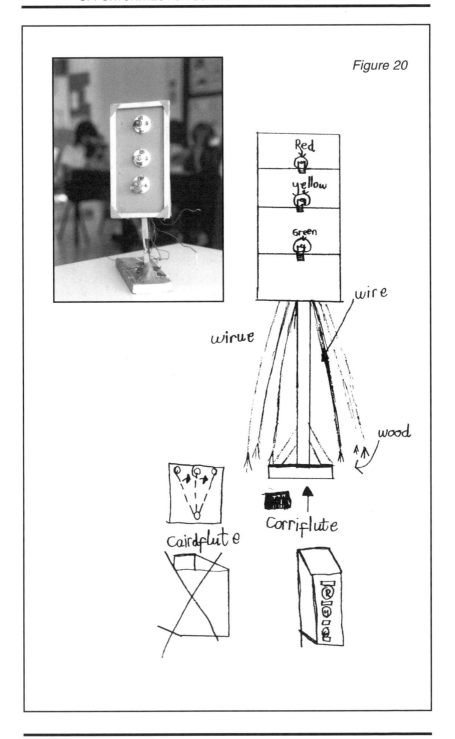

Figure 20

Figure 21

Stefan
Stafan   Traffic light construction   7/8/93

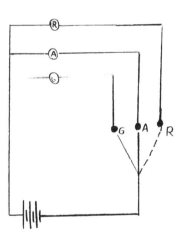

electricity. These matters were usually talked about in a group context when the experience was fresh in the children's experience, so that it could be related to an actual context (Why is it not lighting? Where are the wires connected? Are they touching? Can you see...? What might happen if...? How did you...?)

It was also clear, during these practical sessions, that some children were more capable at manipulating the materials and would therefore end up with a better working product more quickly than others. It is therefore important for the teacher to decide how to share time with groups. In this case, one classroom assistant, available for a special-needs child, worked consistently with that child and her group, in order to help them achieve a satisfactory conclusion; the teacher gave a considerable amount of time to a pair of boys who found it difficult to concentrate on the task unsupported, but who worked well given some teacher input. This inevitably meant that other groups were left to get on, and in most cases this worked out well. However, it was necessary for the teacher to move round and monitor all groups fairly regularly.

Making Traffic Lights

The making and recording of this task took up three weeks. Week 9 was entirely devoted to construction; week 11 to producing a finished set of instructions and circuit diagram. Week 10 however differed according to the stage which particular groups had reached. Groups were largely self-directed throughout this activity, which the children found highly motivating and very satisfying. As with the Public Enquiry activity in the first half term, it provided a climax to the work on the topic.

The task incorporated all the knowledge and skills acquired during the previous weeks, such as designing a circuit; making switches and bulb holders; incorporating these into a circuit; drawing the finished circuit. In addition, children found that they needed to use other technology skills, such as choosing suitable materials for the construction of a support for the lights (many chose to to make a jinks frame for a base, for example), finding ways to tidy up their wiring (some used drinking straws as conduits, for example) and using coloured acetate sheets to create the three coloured lights. There was also an important language dimension in the production of a set of instructions about how to make the traffic lights. Teachers found that even children at year 6 still had difficulty in sequencing the stages, in expressing clearly how to construct something, and in using technical terminology where appropriate. Such diagnosis would again be useful in future activities.

Progression in the production of circuit diagrams was ensured by the requirement that symbols should be used throughout and that straight lines should be used to indicate wiring. The distinction between diagrammatic representation of a circuit, and a drawing of the traffic lights, was emphasised. Examples of the finished diagrams and accounts, along with photographs of the actual constructions, are given in figures 20 to 21. It is interesting to compare the same group's initial circuit drawing (fig. 18) with their final one, to see the progress achieved.

Conclusion

The activity described in this chapter was carried out by two teachers, assisted (and occasionally impeded!) by the author, over a period of a term. The progression in learning is generally evident from the ongoing monitoring which has been described. But how much did we know about the specific learning of individuals? How did we start off each activity? How best are difficult concepts represented? These and other issues are taken up in the following chapters.

CHAPTER 5
LESSONS TO BE LEARNED

The purpose of this chapter is to see what can be learned from the previous three which can help in dealing with some of the difficult aspects of science. By 'difficult' I mean those aspects where teachers generally feel unsure of their own knowledge, as well as those skills specific to science which are often neglected, such as generating testable hypotheses, controlling variables, using appropriate measuring instruments, replicating tests and interpreting evidence. It will focus on getting started; the importance of representing scientific ideas to children; coming to terms with assessment; and will make the assumption that, like other teachers, you are fairly competent at getting children to make and record observations, manipulate simple measuring instruments, and study living things and everyday materials. It will not offer recipes for action, but encourage you to stand back and take a wider-angle view of these major concerns.

Starting points for science learning

Let me start with an activity which is extremely simple, then explain why I chose it.

For this activity, all you need is a longish, smooth thin stick: I have used bread sticks – when doing this in an Italian restaurant – or a metre ruler, slide binder, plastic tube, chopstick, whatever. Balance the stick on the index finger of each hand in such a way that there is more stick overhanging at one end than the other.

Figure 22

Ask the children to predict **what will happen when I move my fingers slowly together until they are touching?**

Try it, before you read on!

Most children (and adults) will predict that the stick will tip over and fall, on the side which is originally overhanging (i.e. to the left in the picture above). Now move your fingers slowly together. You will find that the stick does not overbalance, but that your fingers come to rest touching, at the mid-point of the stick. This happens wherever you put your fingers to start with. You don't believe me?! Go and *really* try it now, if you didn't do so earlier!

Now ask the children to suggest possible reasons why it did not behave as they predicted. You are likely to get suggestions such as;

1. you cheated, you were not moving both fingers.
2. one finger was sticky and wouldn't move.
3. there is more weight on one side than the other.
4. the friction is bigger on one side.
5. the stick is not even or smooth and doesn't slide properly.

You now have several ideas or 'hunches' – what science calls hypotheses – for testing, and you can proceed in various ways, depending on the time, age of the children (or adults; it works just as well!), resources available, size of group. But whether you continue to work with the whole class, or let them break into groups, the next step is to ask them to design a test to investigate one hypothesis. In some cases, this is simple; for example, hypothesis (1) above requires them to 'control' the person whose fingers are used, and to let everyone carry out the test themselves. They will immediately realise that....well, why spoil it! Their experience will prove to them that the first hunch is wrong, or in the language of science, hypothesis (1) can be rejected.

In other cases, this is more difficult; for example in hypothesis (4), how do they go about measuring friction? It may be necessary for you as teacher to choose or match appropriate hypotheses to those who can manage them. But in most cases, children will find a hypothesis at their level. For example, those suggesting the 'more weight' hypothesis (no. 3) will probably want to test their own, and can do this by finding a way of 'weighing' each side simultaneously (probably with two spring balances, or forcemeters). In hypothesis (2), they may decide to make the test fair by using something other than fingers; identical pens or lolly-sticks, for example.

In each case, the activity leads to fresh evidence, which has to be recorded and shared with the group, as the basis for discussion about the hypothesis in question. Children will modify their ideas, and come up with other hypotheses to test. They can happily persist with this activity for up to 45 minutes or more, reporting back and sharing new findings.

You want to know the answer, of course... but that is not the point! If I tell you the answer, there is a likelihood that this will shape the way you present the investigation, ruling in some hypotheses and ruling out others because they are patently leading down blind alleys. You may be tempted to tell them the answer because that is what you see as the purpose; getting the 'right answer' as quickly as possible. But **'being scientific' actually _starts_ when you get stuck and don't know**. By telling answers, perhaps what you are teaching the children is that you are the expert, you know the answers, and that investigation is just something to fill in time and make things fun. They are learning perhaps that you are a 'real' Scientist (with a capital S!) because you know; they are just going through the motions, pretending. Science, they will begin to think, is 'just' a lot of boring facts dressed up in jargon; they copy it down and learn it, but it doesn't mean much. It's important, (everybody says science is important) but they don't know why...and most of them will never be 'real' Scientists...

So, no answer this time! Although that doesn't mean teachers should never provide factual answers. We'll come back to that. But why did I use this activity as an exemplar of doing science?

First of all, 'real' science is about tackling questions or problems which have significance, which people want answers to. When children are faced with something that confounds their experience, they are immediately curious, and want to know more; things are not as they seem, what's going on? So 'surprises' of this kind invariably arouse their interest, which is a crucial starting point.

Second, there are all kinds of possible 'theories' which they can come up with, at all levels of complexity. Some may be amusing, like the cheating teacher or sticky fingers, and easily 'disproved'; both of these introduce in a simple way the idea of a 'fair test'. Thirdly, because the materials and their theories are simple, it is not threatening for them to discuss their ideas. This presents you with an ideal opportunity to explore their schema, by getting them to talk about 'what they mean by' words like force, weight, friction, sticky, as they use them; or such propositions as 'you're not moving that finger'.

Fourth, you can extend the investigation in all kinds of ways, according to their interest and level of understanding. For example, they can make it quantitative by using two spring balances to take measurements of weight at different positions along the rod (I found that a pole-vault pole is good for this, but any long, rigid pole will do): this leads to notions of being systematic (where to take the readings?), keeping records (what to write down, and how?) and looking for patterns in their results (as we gradually moved the hook of one balance nearer to one end whilst keeping the other one fixed, we noticed that...). All these activities provide you with opportunities to assess their level of development in Investigation Skills (AT1).

Fifth, it is a quick and convenient activity to get them to collaborate in groups. Sixth, their conclusions have applications and explanatory power in real life; like for carrying ladders and other things on their shoulders, for explaining balancing and balances, for further work on centre of gravity. It is not a culture-dependent activity; everyone everywhere deals with carrying and balancing. It will probably leave them grappling in their heads with confused ideas about forces, weight and friction: but far better that they engage with these ideas and ask questions than simply write down an answer that they don't really understand. As noted above, **'Being stuck' is usually the *starting point* for science learning**. Children who can observe a teacher engaging, grappling with a real science problem, find it difficult to resist joining in and grappling with it, if only to get to the answer before you do!

And last of all, they can easily try it out at home on their parents or friends, which communicates what they are doing at school in science, and involves mum and dad and others; an opportunity to demonstrate and articulate their developing ideas and science learning.

Taking this example as a strategy for starting off, there are other areas in which the same strategy can be used. What you have to do is set up simple situations in which the experience of their senses conflicts with what they expect. Here are a few examples.

In investigating living things (AT2), children can be provided with unusual fruits and asked to predict what they are like inside, before cutting them open (what do you think the flesh will be like? Will it have pips? Why do you think so?). Most big supermarkets and indian/african/chinese greengrocers will have at least half a dozen to choose from. But don't stop at cutting them open and drawing; the science starts when they try to explain why they predicted what they did (on what evidence?) which leads them to generate hypotheses about fruits which can be tested (it had the same skin and size as an apple, so I thought that...). For example, they will want to associate ideas about thickness of skin, sweetness, hardness, number and size of seeds....the range of hypotheses is endless.

As an example from AT3, when investigating the properties of materials, put some common materials in the deep freezer before asking them to predict their properties. If you 'freeze' cooking oil, grapes, plasticine, etc (in freezer bags, so there is no tell-tale ice on them) they will find them behaving in ways they have not encountered, and again will have to hypothesise about why cooling has this effect. You can then ask them to predict 'what might happen to....if we freeze it?'

When beginning work on water, you can fill a glass to the brim and ask them to predict how many coins you could add to the glass before it overflows.

You might be surprised yourself at the results! You can try it out in the pub quite easily, and become the focus of great curiosity. Of course, there is a lot of science to be got from this in terms of fair testing; level

Figure 23

of water, size of coins, height of drop, steadiness of table, (steadiness of hand...)

And whatever the topic, it is always possible to start with a sorting activity. You can use objects themselves (fruits, stones, leaves, fabrics, metals/non-metals) or pictures on cards (animals, machines, locations) and ask groups to sort them, in whatever way they like. This immediately demands that they decide on categories for sorting: which tells you a great deal about their existing concepts. For example, they may sort animals by number of legs, or wild/domestic, or carnivore/herbivores. You can then impose other ways of sorting, to 'test' their knowledge of other categories; 'sort them into three groups' or 'sort them according to how they move'. This will give you lots of information about what they know and think they know, what interests them, and will be helpful in deciding on subsequent activities.

In introducing children to the idea of complete circuits in electricity (AT4), children usually don't need much motivation as they love playing with batteries, wires and buzzers. However, there is a risk that they associate electrical conductivity only with wires; so you might start by producing a circuit to make a buzz in a different way, perhaps via a tubular steel chair, table leg, a picture frame, a coathanger, or using a floor switch they step on as they enter the classroom. If you can conceal how it is happening (the coathanger inside a coat?) all the better to provoke curiosity and hence hypothesising.

Figure 24

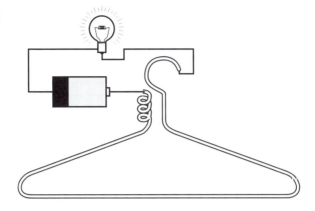

This should provoke all kinds of ideas about what else can conduct, and other ways of completing the circuit.

Children (unlike many adults) don't have too many hangups about electricity in terms of safety and 'getting a shock', as they often cannot distinguish the idea of 'weak' (low voltage) dry batteries from 'dangerous' (high voltage) mains electricity. This has its advantages (they are not inhibited from experimenting) and its drawbacks (they may be tempted to try unsafe experiments outside school, with mains sockets). Hence the importance of introducing ideas of 'being safe' at an early stage. Looking (literally) at electric kettles, irons, lightbulbs and hairdryers, knowing about mains electricity and its force, is essential. Making a short circuit with a battery and wires, either by accident or intentionally, and feeling the heat generated, is a powerful experience for young children, as is making and demonstrating how a fuse works. Let them 'see' and 'feel' that they can't light a 100 watt bulb with an ordinary dry battery, even if it's Duracell! And above all, let them talk about their experiences with electricity in everyday life; things they've seen and felt and wondered about (lightning; shocks from cars and stairrails in department stores; batteries going dead; things going wrong with appliances; 'how does a TV work?') to get a sense of what they need to learn about.

Representing science knowledge to children.

'How to get started' is one thing which worries most inexperienced teachers of science; and 'what to do when you don't know the answer' is another. It is perhaps the main factor in discouraging trainees and new teachers from even trying to teach certain areas of science. So if you've been unable to avoid having a go at Friction or Electricity or Reproduction, what do you do if you don't feel sure of your own knowledge of the subject matter?

An important first step is to be clear about what you need to know and what you don't need to know, in relation to science content. It would be safe to say that you will certainly have no problems if you did GCSE science and feel relatively confident about the main topic areas and concepts. A good first step, if you're not sure, would be to borrow a set of GCSE science texts, and to go through the contents pages, listing the topics you feel unsure about. Having made a (short!) list, you might then go to the sections on these topics, and read a little; you will probably find that you understand a lot more than you think, and that it is only terminology (*is that all it means?!*) that has been putting you off. A great deal of your lack of knowledge can be remedied in this way; and even more important, the reading will have reassured you that you now know where you can find out about Forces, for example, if you need to. Most of these ideas you will have come across before; the 'understanding' that you think you lack will come back quickly, in most cases.

But everybody has blind-spots, areas of knowledge that they are sure they never understood. In science, some of these common areas of difficulty have been covered by texts written specifically for primary teachers, such as the NCC publications relating to Forces, Electricity, Energy. In our own courses, we often tackle these through peer teaching; asking students who do understand the concepts to explain them to those who don't. This works in many ways; partly because the 'teacher' knows what it is the 'learner' needs, and because they 'speak the same language' as a consequence. It also helps both teacher and learner, by having to articulate and be questioned about difficult ideas.

Peer teaching can work in school just as well as in a pre-service training course. The real difficulties are admitting you don't know in the first place, and finding time when you and the Science Co-ordinator or another colleague can get together to work on it. Collaborative teaching (team-teaching) is another effective way of learning not only the science concepts but the kinds of questions children will ask, so that you can be prepared for them in future. More attention is given to this in chapter 7.

However, research on teachers' subject matter knowledge has suggested that it is not simply knowing the science that is important in teaching it effectively. Researchers have distinguished between three kinds of teachers' knowledge, as follows:

- Subject Matter Knowledge (SMK; i.e. in this case, science content)
- Pedagogical Knowledge (PK; i.e. general teaching methods)
- Pedagogical Content Knowledge (PCK)

and it is the third of these which is generally regarded as being crucial. So what exactly is PCK?

PCK can best be thought of as **knowledge about the most effective ways to represent science ideas to children** so that they can be learned. It is not simply general ideas about teaching science (using hands-on activities, for example, to make science more interesting and meaningful) but specific knowledge about effective ways of representing the idea of Energy or a Fair Test. So whereas Pedagogical Knowledge can be thought of as general and relevant to teaching most subjects, PCK is specific to science and to the ideas within the science curriculum.

As an example, the use of analogy is a generally useful teaching method. But the specific analogies that are valuable in representing specific science ideas is an important form of PCK. For instance, the idea of water flowing over a waterwheel as an analogy for current electricity often helps children see why electricity is not 'used up' as it passes through a lightbulb.

Other useful analogies for electrical current are water flowing through a pipe (where squeezing the pipe illustrates resistance) or the chain of a bicycle when pedalled, which illustrates that current flows round in a circuit, and that it is not the electricity in the 'chain' that gets 'used up',

Figure 25

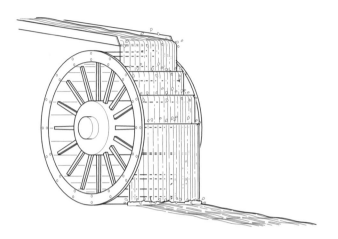

but the energy in the battery (in this case, the pedaller) which is changed into something else. Summers and his co-researchers (1996) have shown that use of such analogies significantly improves the learning of both children and teachers.

There are many valuable analogies which can serve as representations of science ideas. Bruner introduced the notion of enactive, iconic and symbolic representations of concepts (Bruner, 1960) and the teaching of science has developed a multitude of ways in which each of these can be used. Amongst them are familiar ideas such as diagrams (the good old Water Cycle, which every teacher knows!), models (the Smoking Machine, described in the previous chapter, to show how tar collects in the lungs), and simulations (a circle of children holding and squeezing hands, to show the flow of a current round a circuit).

How then do you 'get' Pedagogical Content Knowledge? It is probably useful, first, to see professional development as a career-long thing; you will increase your repertoire gradually as you go along in different ways, and can't be expected to have a bagful of ideal representations before you start out. You can also find examples in the many journals, TV series and INSET videos available to teachers, which means you have to find time to read and look at them. You can also draw on the expertise of your Science Co-ordinator and other colleagues; and perhaps most easily overlooked is the expertise and insights of children themselves in representing ideas they have grasped.

But perhaps the most useful, yet under-used help with representing science ideas is the textbook or scheme. We are one of the few countries in the world – Australia is about the only other – which chooses not to teach science from some form of mandated text. Schools may own a scheme or two – the evidence however is that they have bought less and less in recent years – and the Teacher's Guide may be used as a source of ideas for activities. But there is growing evidence to show that not only do teachers rarely use Pupils' Books with children in science, but also that trainees are often inadvertently discouraged from using published material, through being encouraged to differentiate and hence to produce their own materials. How to use text more effectively with children will be the focus of the next chapter.

Assessing children's science

In many books on science teaching, assessment and evaluation has often come in the last chapter, almost as an afterthought. However, since standard testing in science became a legal requirement in England and Wales (though not in Scotland), teachers are aware of the need to carry out – and therefore prepare children for – the SATs. Simultaneously, awareness of the need to know what children know, in order to match activities to their needs, has developed through the spread of constructivist ideas and research. Few teachers will finish their training these days without having considered the SPACE Reports on children's science ideas, or the ideas of Vygotsky and Bruner relating to diagnostic learning and mediation through scaffolding. All of which means that assessment of children has to be planned into teaching programmes.

The National Curriculum and its concomitant demands for assigning levels at the end of each Key Stage have focused attention on the Formative-Summative dichotomy in assessment. On the one hand, we have externally-set SATs at the end of Key Stage 2; on the other, we have the ongoing process of teacher assessment, which demands a plethora of techniques for gathering evidence about children's science learning, for recording it and for doing all this with the minimum of stress. The rationale for summative assessment and levels is to do with raising standards, comparing school performance and parental choice:

the arguments of the market. For teacher assessment, however, the rationale is about feedback and improved individual performance. And the whole merry-go-round is driven by a widespread willingness amongst teachers, parents and administrators to believe that it works: that if we assess children they will do better, and so will their schools.

Being a scientist, however, means asking questions; and this assumption is worth questioning, mainly because the process of assessment now imposed on teachers in science is extremely difficult to sustain. It is difficult for many reasons. First, assessing *individual* attainment or understanding when children are deliberately being asked to work *collaboratively* in order to learn effectively, is bound to present problems. Second, there is increasing evidence to show that children's language is often a barrier to adequate communication of science ideas, and that this is more serious with young children and those learning in a second language. How do we know that what they say is what they understand? How do we know what they understand by the questions we put to them?

And third, assessing children doing science is very time-consuming, simply because the process of doing science is time-consuming. If I want to observe children working in groups (to find out for example what part an individual has played in planning an investigation, or if she sees the need for controlling variables or repeating a measurement), I must spend a long time with the group; or risk missing the crucial stage of the activity. Various authors have attempted to provide help with this process, such as the publications of the STAR project based in Leicester, to the publications of SEAC and more recently SCAA itself, with the publications on exemplification of standards, and to books from those with long experience of assessing science performance such as Qualter (1996). What all these confirm is that science poses problems for assessment which do not exist in the other core subjects.

It is hardly surprising, therefore, that little evidence has been found that teachers get better at matching activities to children's needs on the basis of formative assessment. We feel intuitively that monitoring, assessing and then matching tasks must improve learning; but an alternative view has recently been expressed in a report comparing science achievement in primary schools in different countries, by

Reynolds (1996). This view starts from evidence that the range of attainment in English classrooms is greater than in other countries, and suggests that attempts to differentiate actually accentuate this range, by holding back the lower attainers and helping the high attainers to move on more quickly. What many countries do, explains Reynolds, is to hold back low achievers to repeat a year in order to narrow the range of attainment in a class, so that more whole class teaching (and therefore less differentiated assessment of individual attainment) can be used.

This book is not going to come down on one side or other of this argument. There is clearly a strong case for the efficacy of individual monitoring of learning in science; if it can be carried out within the constraints of the real-world primary classroom in which you work. It is incontestable that planning effective science teaching is helped by finding out what children already know, believe and can do; and that this 'finding out' is one key aspect of formative assessment, underpinning all currently accepted theories of how children learn. The questions about assessment are not about whether, but about how, and if.

But it could be argued that in the English system we have chosen an ideal but relatively impractical route to assessing children; whilst many other countries have opted for a less than ideal but more feasible process, based on frequent testing. The issue cannot be resolved, either, by simply concerning ourselves with the methods: the historical, social and political context in which assessment is used carries great power to determine whether or not a system will be effective in achieving its aims. For example, the importance attached by parents to the assessment at the end of primary school depends crucially on the post-primary choices available to children and parents, in terms of such things as the status of schools (selective schools, private schools, segregated schools). The less significant these issues (for example in Germany) the less important terminal assessment becomes. In many African states, on the other hand, where there is no post-primary education whatsoever for the majority of children, assessment in the primary school phase becomes crucial and intensely competitive, and dominates teaching as a consequence.

For there is no doubt that assessment, where it exists, drives teaching and learning. Secondary science teachers will admit that much of the note-taking and copying from texts which goes on in GCSE science lessons is intended to help students be able to revise for exams. This inevitably affects primary teaching in the same way (most primary trainees come to training with negative perceptions of school science!) and can present real dilemmas for teachers. One example from the activities described in the previous chapter will suffice.

In, 1993 teachers knew that there would be SAT questions about wild flowers and where they grow; for example, dandelions grow in meadows...but what happens when on your environmental study, your children find dandelions in the woods but not in the meadows? The kind of rigorous investigation that might resolve this (looking at various woods and meadows at different times, for example, to establish probabilities and trends) is not feasible in the the primary science curri-culum. The great temptation is to tell them the 'right' answer, i.e. the answer that they will need to give in order to gain the marks (even though, as in this case, the answer is too simplistic to be 'right' in any real sense). Gradually, resorting to such behaviour will erode children's and teachers perception of the value of investigative science; we will be back to the 'what's supposed to happen?' syndrome of so much point-less practical science, where the actual result obtained by investigation becomes increasingly perceived as insignificant, and all that matters is the perception of science-as-facts-and-figures which is anathema to so many students.

Does assessment have to be like this? The theoretical answer is 'no'; we can choose to have criteria for our formative assessment which focus on what we want children to learn, and we can communicate these clearly and consistently to children, so that they know what they have to achieve in order to succeed and gain teacher's approval. For example, we can, in theory, reward any of the following, which are related to good science practice:

- children's demonstration of curiosity leading to original observations;

- persistence with a task in order to obtain a complete set of results;

- choice of appropriate instruments;
- consistent and accurate measurements;
- improvising of techniques;
- choice of an appropriate way of recording observations;
- oral communication in clear, unambiguous language using appropriate technical terms;
- willingness to generate and explore alternative hunches.

We can stop rewarding neatness, conformity to a preferred style and layout, accurate spelling, and behaviour which may not be productive and on-task but which is conforming, polite and non-disruptive.

To do these in practice, however, means taking account of the likely responses of all the audiences of our assessment, which include not only the children but parents, other teachers, Head, governors, Ofsted... and many of the 'positive' criteria are invisible to record and hard to display, unlike the neat third-drafts of accounts of investigations, double-underlined and triple-mounted for Parent's Evening. Unlike the quiet, busy classes all Heads and Inspectors like to see.

For several years, I had the privilege to work in Kenya with a brilliant teacher of science in a rural primary school. He put into practice all the ideas he believed in; focusing on problems identified by the children themselves (such as chopping wood, keeping fish, eradicating insect pests, preventing teenage pregnancies...) encouraging open-ended investigation and planning in groups; daily brain-storming of where children had got to in their investigations, what the problem was, ideas of how to proceed; consulting local experts such as herbalists; and detailed recording and sense-making. The work of his class in science was quite unstructured, often noisy, seemingly slow and inconclusive; and unlike in other schools, did not involve regular going-over of past exam papers. At the end of their primary school career, like all other Kenyan children, his class took the primary leaving exam in English, Maths and Science for entry into secondary school, and always performed better than any other school around, not only in science but particularly in English and Maths.

The lesson I draw from this example is that teachers must make their own minds up about where they stand on assessment; and that it takes courage to buck the pressure to conform, but can be well worth it, and can produce better results even in the 'hard' terms of standard tests. The operation of the National Curriculum legally compels us to conform to specific requirements in terms of recording progress and reporting this to parents; but there is still considerable scope for schools and teachers to manage assessment in their own way. Our own evaluation of assessment and reporting procedures in one LEA, discussed in more detail in chapter 7, confirms that in reporting to parents, least attention is paid to children's skill and concept learning in science, while most is given to behaviour and attitudes, assessment of which is not required by law. What parents often want is to know what their child is *going to do* in science so that they can help, rather than what they have already done and finished with. Some may even want to assist in the assessment process by looking after groups while you focus on individuals. Many things can work to help you use authentic methods to assess. On the other hand, teachers who work with second-language learners, with little parental help, frustrated at their having to assess children in a language over which they have limited command, and criticised unjustly by inspection reports and the media, well understand the need to maintain a scepticism about the purposes and interpretation of assessment results.

CHAPTER 6
USING PUBLISHED MATERIALS EFFECTIVELY

Problems with the use of published science materials

The paradox relating to the use of science text materials in primary schools is that on the one hand they are seen as central to the process of teaching and learning science, whilst on the other hand there is little evidence of their effective use by children. This is true not only in second-language (L2) and developing-country contexts but also in countries such as the USA and UK where science teaching in primary schools has been fostered for several decades, and where masses of published material is available. In the USA, for example, Shulman (1987) concluded that most teaching is initiated by some form of text; yet Ball et al. (1988) demonstrated that teachers in training were discouraged from using the set texts, being instead encouraged to develop their own materials. In UK teacher education institutions as in the USA, trainees are not extensively taught or encouraged to use existing text material in science: instead we stress the importance of differentiation and therefore are more likely to encourage trainees to develop and use materials (often worksheets) of their own construction.

Why should we have all this professionally-produced and expensive material and not really use it? The answer relates partly to the nature of the materials themselves, partly to our ideas about good teaching. Children tend to learn to read and use text material largely through the use of story-books, which are narrative in structure. Science books, on the other hand, are expository: they do not simply tell a story, but set out information, ask questions and give instructions, often by the use of diagrams and illustrations as well as words, and often using all kinds of design and format conventions (boxes, highlights, variable columns). The language of expository text uses vocabulary and connectives which children have not encountered in story-books. This new kind of text material can pose problems for many children, particularly those learning in a second language: the text can become a source of difficulty, rather than a source of help.

The most extensive evidence relating to science texts and L2 learners comes from the research of the Threshold Project in Southern Africa (Macdonald 1990a, 1990b; van Rooyen, 1990; Langhan, 1993). Again, text materials are here shown to be available but ineffectively used: and the paradox is partly explained in terms of the quality of the materials themselves.

The Threshold team exposed difficulties of various kinds with science text material, including;

- problems of children's comprehension, the demands of the text being too great for the levels of language development of the children using them;

- problems of mismatch between teaching styles implied in the text and those normally adopted by teachers in the contexts studied;

- problems arising from teachers' own perceptions and understanding of science and the subject matter content of the text.

The project found that teachers did not follow the demands of the text material, but rather interpreted the materials in their own way, maintaining traditional teaching styles and forms of interaction that their pupils were used to, such as teacher talk, closed questions and pupil copying of notes. When project staff re-designed and re-tested children on the new materials in the light of pupils' previous difficulties, modest learning gains were noted. However, the project also reported that teacher adaptation of materials often meant that they no longer provided the pupils with any text resources to use themselves.

The same situation has been observed in our research in both London and Kenya. In London, teachers in classes where over 95% of children were L2 speakers justified not using text material with the children on the grounds that such materials did not meet children's needs, demanded language abilities beyond the children's levels of development, didn't foster group discussion, demanded too much teacher mediation and were not written by authors with an understanding of the specific mother-tongue context (Peacock, 1995a). Simplified worksheets were often adapted from sections of the schemes, and teachers often suggested ways in which texts could be improved, in terms of vocabulary, use of graphics and of different cultural expressions and meanings. However, teachers rarely if ever referred to the importance of children needing to be able to deal with published expository text, for example, in order that they would be better able to cope on entry into secondary school, where use of such textbooks was more widely prescribed.

In Kenya, ongoing study in primary schools in urban Nairobi and rural Western Kenya suggests that provision of mandatory texts varies in quantity from school to school, but that the text is still used extensively in classrooms. However, the main use of the pupils' text by teachers is to require pupils to copy material from textbook to notebook (Murila, 1996). Kenyan teachers were reluctant to explain new words from the text, partly because of pressure to complete the overloaded syllabus and partly as a consequence of their own lack of understanding, both of the text and of the subject matter represented in it. These observations echo the findings of most other research in Africa on primary teachers' science misconceptions (for example Rollnick and Rutherford, 1990). They also emphasise the need for primary teacher trainers to understand and teach not only the subject matter content of science but also, and crucially, those reading, writing and visual literacy skills appropriate to the use of expository text, with particular emphasis on children learning in a second language.

Ways of looking at science text material

Several strands of research exist in relation to the analysis of text material in science, all of which shed some light on the paradox of their under-use in classrooms.

Linguistic analyses of science texts has been largely focused on the secondary phase of schooling and on first-language contexts. Evidence relating to primary schools and L2 contexts is summarised in Peacock (1995b). Briefly, the research which exists suggests that:

- readability measures are too crude an indicator of pupils' potential difficulties;

- texts used during the stage of transition from mother-tongue to L2 medium of instruction are crucial, and yet are often not written with progression in mind, for example in terms of the different demands made by science texts and earlier L2 language schemes;

- texts are usually written by first-language speakers as if for first language speakers, and take no account of the language differences of minorities. For example, the fact that some children in the UK speak a language at home (such as Sylheti-speaking children of Bangladeshi parents) which is not written down and therefore does not have any books.

The literature on visual literacy suggests that the actual function of graphic material (illustrations, diagrams, photographs, tables) in science texts is much more complex than imagined and is often therefore ineffective. Illustrations do not always motivate and arouse interest, as assumed, but sometimes distract attention: being able to read the language of the text and having previous experience of graphics is an essential precursor to visual literacy. In developing countries, rural children have much greater difficulty interpreting diagrams and pictures than do urban children. The findings are again summarised in an earlier article (Peacock, 1995b).

Science books can also be analysed in terms of their structure and format. They tend to use graphics and text for a complex range of functions, and so are also 'more literate' in terms of the 'deep structure' demands made (Cummins, 1983; Vachon and Heaney, 1991; Gilbert, 1989; Hyltenstam and Stroud, 1993). Biber (1991) has stressed the dangers of simplification by teachers: trying to make the text easier can impoverish the concepts being taught, and does not necessarily improve comprehension of science content. But there is very little research on this important area of understanding.

Dowling (1995) has analysed texts from a sociological standpoint, in terms of how they 're-produce' activities. His study of primary school textbooks in mathematics introduces concepts such as Textual Strategies, Setting, Message and Voice to discuss how texts work, and concludes that such texts create a 'myth of participation'. What this means is that since the teaching method is not usually made explicit in the textbook, 'knowing what to do' with the text has to come from either the learner or the classroom setting. For example when the 'voice' of the text says 'collect some different materials...' the pupil has to decide somehow whether *actually* to follow this message (is this what my teacher wants me to do?) or simply to read on to where the text provides answers or information. Yet learning the science that the text intends is often dependent on collecting the materials and observing them: so that if the classroom setting prevents the pupil from doing this, participation (and therefore learning) doesn't take place.

This seems to happen in many classrooms, in both developed and developing countries. For example, evaluations of the innovatory 'Spider's Place' science materials in comic-book and video form in South Africa (Perold and Bahr, 1993) showed that, where the teachers and trainers implementing the new materials possessed a limited repertoire of teaching strategies, the materials were not used as intended, and the participation implied by the messages in the text did not take place. Testing of new primary science texts for L2 learners which puts materials directly into children's hands is also suggesting that, without the presence of a teacher, pupils do not follow the 'message' of the text, and therefore do not acquire what the text intended (Francis, 1996). Instead, whilst the text implies one set of teaching methods, pupils often make sense of it by reference to another and very different web of meaning determined by their own culturally-determined knowledge of what is expected in the classroom (Kouladis and Tsatsaroni, 1996).

This problem is often made worse by the fact that formal teaching of reading tends to stop just at the stage when pupils are beginning to meet and use expository texts for the first time. Evidence suggests that many children are never taught the new strategies needed for reading and making sense of expository text material. Roth (1985) has suggested that both good and poor readers have difficulty learning from science texts for this reason. Texts rarely talk to their pupil audience about the metacognitive ('learning how to learn') aspects of what they are trying to achieve; exceptions are the 'Spider's Place' series already referred to, which has a section at the beginning of the pupil's comic (the text) which addresses both content and appropriate learning strategies, and the Nuffield Primary Science materials, which overtly draw attention to constructivist ideas of concept development throughout each text.

What does this mean for teachers?

In relation to pupils' comprehension, teachers (especially of L2 pupils) are often aware of the problems inherent in commercially produced texts, and tend to respond to these difficulties by not using such text material, choosing instead to develop their own materials, even though these materials may impoverish content and not improve comprehension, and even though their pupils will sooner or later need text-processing skills.

Texts for pupils have within them implicit messages about teaching which, whilst they may be made explicit elsewhere (for example in Teachers' Guides) are only rarely made explicit to pupils. These implicit messages are often at odds with the methods adopted by teachers in their classrooms. Left to themselves, L2 children (especially in developing country contexts) do not appear to follow the task messages in a text, even when these are explicit, and require the individual help of a teacher to make sense of what the text demands. Yet most teachers stick to their familiar ways of teaching, rather than those implicit in the text, so that a 'myth of participation' is perpetrated by the text and often reinforced by the teacher.

Though considerable research has often gone into science content at the text development stage (for example the SPACE research project and the subsequent Nuffield Primary Science materials, 1993), the role of text seems to be seen by many authors, designers and publishers as 'simply' to re-present this knowledge in an attractive format. Yet whilst pupils' texts ought to represent science knowledge to children in the best possible way (because they are written by experts), they are rarely used in classrooms in the way the authors intended.

Part of this problem may be connected to training. Primary teachers of L2 children do not seem to be given much pre-service training on the use of expository text. Rather, there is evidence that training programmes discourage the use of text, and encourage the development of personal contextualised materials by trainees. However, pupils will sooner or later need to be able to process expository text if their learning is to continue, and therefore teachers must themselves be able to teach these text-processing skills. A crucial question is; what do teachers need to know about using science textbooks, in particular with L2 children? To try to answer this, it is helpful to look at some specific examples of published materials. Some of these are from developing-country contexts, because they highlight the L2 difficulties and because there is so little similar research evidence from the UK.

Analysis of examples of text material.

Extracts from four texts will be analysed in terms of what they attempt to represent, and how children responded to them, leading to suggestions about how to help teachers to use text more effectively.

Example A (fig. 26) is the trial version of a lesson on Evaporation from a commercially-published South African text. The trial version was given to a class of children in the absence of their teacher; they were then observed using it and interviewed afterwards.

The title of p.1 does not indicate what the substantive science focus is. The first instruction is in three stages ('make a puddle...', 'draw...a line...' and 'go back later...') and incorporates illustrations to represent the activities visually. The page ends with a question requiring the children to suggest an explanation of how the water disappeared. Adjacent is a box containing an illustration and a question relating to their general knowledge about drying washing on a line.

From a linguistic view, a crucial problem on this page was the word 'puddle', which most children did not know. Only when they were given a bottle of water and questioned about the first illustration were they able to understand what the first step required of them. Two-thirds of the children did not see the three 'starred' steps as being related in a logical progression. Hence they stopped for further instruction after step 1.

Figure 26

Going Up?

Watch some water disappear

☆ Make a puddle outside on some tar or cement.

☆ Draw or scratch a line where the edge of the water is.

☆ Go back later and see what happened.

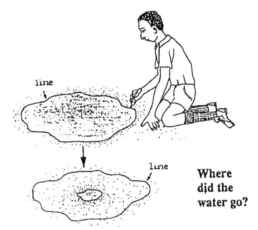

line

line

Where did the water go?

Do it again in a saucer

make a mark

after a few days

mark

Where did the water go?

What happens to the water when wet clothes are hung up to dry?

Maybe some of the puddle-water soaked into the ground. This can't happen in a saucer or a bowl. It must have got out another way.

Figure 26 continued

Escaping water

Water is made up of tiny bits or particles that are too small to see. At the surface of the puddle or saucer of water some of these particles move out of the water and go into the air. You can no longer see them - they are invisible. we call these invisible particles water vapour. Water vapour is invisible, it is dry and it is a gas (not a liquid). You cannot drink it and it cannot make you wet.

The liquid water in the saucer evaporated: it became a vapour. (Vapour is another word for gas.)

Challenge

Work out some ways that you could make water evaporate more quickly. If you use two saucers (or two pots) that are the same, you could make the water evaporate quickly in one. Then you could compare it with the other to see if you were successful.

Slow down evaporation and save water

Most of South Africa is a dry country - we need to use our water carefully. One way is by mulching.

After planting your vegetables or young trees, cover the bare soil around them. You can use cut grass, leaves, or torn-up paper.

In very dry areas you can even pack stones around young plants.

Mulch:

☆ keeps the soil cool in summer.

☆ stops water evaporating quickly.

☆ makes it difficult for weeds to grow.

☆ begins to decompose and feed the soil.

Water is evaporating all the time - from ponds, rivers, dams, puddles and the sea.

Play a trick on a friend

What kind of water cannot make you wet and cannot be drunk? I don't know

Water vapour. It's a gas! oh!

The visuals presented several problems. Most children thought that the boy was drawing a picture of the puddle; the dots in the illustration were thought of as sand; the two arrows indicating 'line' were correctly interpreted, but the function of the central arrow (indicating progression to the later stage of the activity) was not understood by anyone. And in the second illustration, presumably by analogy with the first, pupils interpreted the arrows as requiring them to write 'make a mark' on the saucer. Some wrote 'make a mark after a few days', eliding the two separate instructions in the illustration. Most children ignored the boxed question (bottom left).

On p.2, pupils were presented with information about evaporation in written form. Only a minority understood the title before they started reading. Pupils were unable to link 'tiny bits' with 'particles', and 'you can no longer see them' with 'invisible'; a minority could point to the 'surface' of a puddle. Some could identify names of liquids but could not explain what a liquid was, even though the definition is implicit in the last sentence of the first paragraph. Unlike the box on p. 1, children's attention was drawn to the cartoon on p.2 even before they read the preceding text, perhaps suggesting that they were familiar with this 'comic' format (Francis, 1996).

The children's responses illustrate the 'myth of participation' concept quite clearly. They had to be prompted to act at each stage, even though the instructions are clearly practical and in the imperative. However, there is a practical illogicality in step three ('go back later...') since they cannot actually do this during the same lesson; so they have to decide; should they carry on with the rest of the page or not? Here, it is a question of whose 'voice' they are hearing: pupils did not act until the 'teacher' directly reinforced the message of the text. The voice in the text is not the teacher's voice. Yet the observer reported their enthusiasm to try it out at home and observe what would happen.

Example B is from the text currently used in Kenyan primary schools in standard 4, the year in which children first undertake all their lessons in English (Kendall, 1990). Figures 27 and 28 illustrate two pages from the unit on 'How animals and plants move and feed'. The first, p. 36, uses text and an illustration to ask questions about the feeding of 'farm' animals ('what are they feeding on?', 'what is the boy giving the rabbits?'), incorporating also some simple substantive science knowledge which pupils are likely to possess already ('Animals must feed. If they do not receive food they will die'). The second (p. 39) again requires pupils to use a picture to answer questions about animal movement, although it is not made clear how they should demonstrate how animals move (writing? orally? by actions?). In addition, p. 39 requires children to 'look outside' or 'take a short walk', and then to complete a table showing how observed animals move, a logistically and practically difficult task during a primary school science lesson.

Neither page represents (i.e. *tells* the pupils) any new substantive science knowledge; in both cases, the text 'message' is that children will acquire new knowledge by carrying out the tasks of observation required.

The researcher observed the activities being taught to classes of children by their own teacher. In both cases the teacher's voice over-ruled the messages of the text, so that in fact any new knowledge the children acquired was presented by the teacher on the blackboard for copying. The myth of participation in the text activity is reinforced, and the text's main message (you can learn by observing and finding out for yourself) is rendered redundant. The lessons were taught mostly in English by most teachers; pupils answered some questions in Kiswahili because they did not know the English names of plants and animals, for example. Teachers usually accepted these vernacular words without providing the English equivalent. Most of the lessons consisted of teacher talk; all teachers worked from their own notes rather than the text. Most did not expect the children to open the text until after the teacher had 'finished teaching'; often the teacher copied notes onto the blackboard and asked pupils to copy them quickly before the end of the lesson. Teachers justified this because most classes and pupils did not normally have textbooks, having to share, being too noisy when they used textbooks, or because texts were 'not detailed enough'; their notes tended to use technical terms not used in the text. In the one class where the teacher used the text, she made the pupils read aloud in turns and occasionally stopped to explain something or to ask a question. Pupils were commended for correct answers.

Figure 27

UNIT 6

How animals and plants move and feed

How animals feed

Plants and animals are living things. Look at this picture of a farm. What is the mother doing? What is the boy doing? What are the girls doing?

Fig. 6.1 Some activities that go on in a farm

The animals on this farm are being given food. Look at other parts of the picture. Can you see other animals feeding? Name six other animals feeding. What are they feeding on?

Animals must feed. If they do not receive food they will die.

Which plants can you see in the picture? Point to the tree. Point to the flowers. What is the lady giving the cow? What is the boy giving the rabbits? What is the girl feeding the hens?

The cow, the rabbits and the chickens are receiving plant foods. Which animals in the picture are feeding on other animals?

Figure 28

Movement in animals

Look at this picture. Which animals can you see moving? What movements are they making? How are those movements helping them?

Fig. 6.6   How does each of these animals move?

Activities

Look outside your home or classroom. Take a short walk. Write down the names of 10 animals that you find moving. Write their names in a list, say how they are moving and how their movements help them. Record your answers in a table like the one below.

Name of animal	How is it moving?	How does the movement help it?
1. Elephant		
2. Antelope		
3. Lion		
4.		
5.		
6.		
7.		
8.		
9.		
10.		

It was also clear from talking to pupils that they did not understand some words in English, and that they did not look at the illustrations when reading, even though they could explain, when asked, what was required by the table on p. 39. Subsequent interviews revealed that the teachers were generally not aware of any of this (Murila, 1996).

Example C (fig. 29) is from 'How to become a Great Detective', one of the 'Spider's Place' Comic books from South Africa (Handspring Trust, 1993), also broadcast as a 13-part TV series for schools on SABC. The evaluations of these materials in use found that the most common teaching strategy used by teachers was to ask pupils to read through the text aloud, in turns, and for the teacher then to ask questions about the content of the text (what did Ayanda mix with the vinegar? What colours did Jay and Frankie see?). Rarely if ever were pupils subsequently observed doing a chromatography test themselves, or discussing how Ayanda and Spider made the unfair test fair, even though these are the substantive science messages in the text. However, the evaluation also showed the great popularity of the comics with pupils, who sought to take them home (and even stole them!) in order to read them, which they did without difficulty.

In the text, a narrative structure is used, and the 'voice' is the voice of pupils themselves, using vernacular terms ('awuzwe-ke!') at times. The teaching strategy, which is made clear in the introductory meta-cognitive section, explicitly emphasises pupils themselves using science skills to solve problems. Yet the approach was not transparent to the teachers, who either misunderstood or deliberately ignored the text message, and who often justified this in terms of class size, lack of materials, pressure to complete the syllabus or disruption caused by practical activity (Perold and Bahr, 1993; Peacock and Perold, 1995).

Example D is from the pupils' book on 'Electricity and Magnetism' in the Nuffield Primary Science materials (fig. 30). It has not been evaluated with children as part of this research programme (mainly because we could not find schools in our sample which used the Pupils' Books with children) but will be analysed in terms of the role it is intended to play in presenting subject matter to children.

The Teachers' Handbook stresses that pupils' books are supplementary and not intended as a substitute for practical work and discussion: they

Figure 29

Figure 30

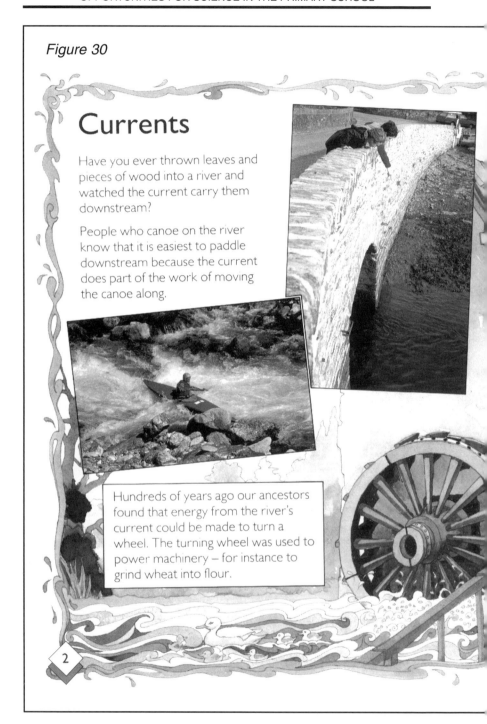

Currents

Have you ever thrown leaves and pieces of wood into a river and watched the current carry them downstream?

People who canoe on the river know that it is easiest to paddle downstream because the current does part of the work of moving the canoe along.

Hundreds of years ago our ancestors found that energy from the river's current could be made to turn a wheel. The turning wheel was used to power machinery – for instance to grind wheat into flour.

Electricity is a bit like the current of a river. In fact, people talk about 'electrical currents'. An electrical current supplies the energy which makes the blades of this mixer move.

Turning this switch on allows an electrical current to flow from the **battery**. If the current is made to flow through a lightbulb, the energy of the current will light the lightbulb. If it flows through an electric buzzer or bell, they will make a noise.

Can you think of some other ways that the energy of an electrical current is used?

3

can be used to 'start small-group activity', to 'arouse curiosity or provoke questions for discussion', as 'links to work in other subject areas' or to 'revise or consolidate classroom teaching'. It is recommended that they are 'used flexibly as an aid for learning' in 'organisation of work that children can do on their own or in small groups' (Nuffield Primary Science, 1993).

In relation to the example spread on 'Currents', the Teachers' Guide to the Electricity and Magnetism unit mentions the concept of current in several places; for example, as new vocabulary to be introduced; as a concept with which children have difficulty; and under 'Background Science', the Guide provides a technical explanation of its meaning in terms of movement of electrical charge, direct and alternating current, mains electricity etc. Children will also presumably do the activities on constructing circuits described in the Teachers' Guide, and their attainment will be assessed through observation and questioning.

In order to decide how and when to use the section on 'Currents' in the pupils' book, therefore, what does a teacher need to decide? Our earlier review of research suggests that a teacher must:

- assess the language level of the text and compare it to that of the pupils';

- consider the illustrations, assess the difficulties and demands they might make on pupils, identify possible ambiguities;

- consider the 'message' of the text, the transparency of task demand (if pupils are to use the text independently) and the extent to which active participation is required by the 'voice' speaking in the text;

- identify the concepts covered and the implied conceptual development, and match this to the level of development of the pupils.

Even a superficial analysis in these terms reveals that in each case the text is likely to present ambiguities which will demand teacher mediation, particularly with L2 children. For example, the concept 'Energy' appears four times in four different parts of the page, but is not explained, nor does it appear in either index or glossary at the end of the book. The diagrams of the mixer and of the bulb connected to the battery are both intended to illustrate flow of current – 'through a lightbulb' as the text says at one point – and yet the illustrations could be taken to show wires actually entering through the glass of the mixer bowl and the bulb in each case. Questions are posed ('Can you think of some other ways in which the energy of an electrical current is used?') but the message about how to answer is not transparent: does this expect an oral or written response, or neither? Clearly, the teacher would need to anticipate these difficulties and assist the reader to get over them in order to use the page effectively and independently.

Implications of the evidence about the role of text and its use

These examples – and they are not exceptional in any way – show that there are many potential obstacles to the effective use of any science text by pupils. The role of text, even within the range of examples provided, is clearly diverse in terms of the intentions of authors. Yet the messages about role and use embedded in these texts are often not transparent; and in some cases, even where the message is clear, it is over-ridden in use by the perceived pressures on teachers and the expectations of pupils. It is also clear from some of the examples analysed that the subject matter knowledge 'covered' by the text can only be learned if pupils participate in the activity demanded by the text; yet such participation is rare, sometimes because it is not logistically possible, sometimes through not being demanded by the teacher.

Pupil use of text is thus likely to need teacher mediation most of the time; and teachers are likely to need help not only with their understanding of the content of text, but also with the analysis of the specific text material they are likely to need to use. This means there must be collaboration, in the training of primary teachers, between trainers responsible for curriculum science programmes and other programmes, such as those in language and literacy; and it suggests that teachers, particularly during initial training, may need more help than they currently receive with how to make best use of existing text. It is likely that such help will also need to be text specific, because the difficulties will differ from one scheme or book to another. Help will also need to be context specific, depending on the age and language background of the children being taught and the teaching styles preferred in schools. Research and development for such training is urgent. It might usefully build on progress made by studies of how children can be helped to interrogate non-fiction text, such as that of the EXEL Project (Wray and Lewis, 1995).

The final question is, should a teacher focus on using existing material in the school, on obtaining better materials, or on producing customised materials? Work to develop better texts is going on, but it is widely accepted that there will never be a 'perfect' textbook which will be appropriate in all the different contexts likely to be encountered. Current financial constraints in UK and in developing countries mean that fewer books are being purchased for schools (in a well-resourced London borough, the most commonly owned schemes were both around 20 years old!). Similarly, there will be always be room for teachers 'hand-made' worksheets and handouts, but most pupils are wary of 'death by worksheet', and it is clear that writing your own can be a minefield leading to difficulties as great or greater for pupils than the ones found in published materials. Children will sooner or later (certainly on entering secondary school) need to be able to use the text effectively, whether or not it is a good one. Teaching them to use it must be preferable to having them torment teachers with their endless 'I don't get it, miss!' It is a good idea to look closely at the text *your* children will use in science when they move up to secondary school, and to analyse it as above, to recognise the skills needed to use it.

CHAPTER 7
OPPORTUNITIES FOR COLLABORATION IN SCIENCE WORK

Collaboration in science work with children is important for the following reasons:

- The predominant view of learning in science is that children construct ideas based on what they already know, modifying these as a consequence of experiences and mediation by teachers and others. In this view, teaching is seen as assisting the performance of learners in constructing new ideas, and is therefore essentially collaborative

- The practical and investigative nature of school science requires children to pool ideas in the planning stage, to divide up practical tasks to make them manageable, and to discuss the meaning of the results obtained

- The key principles underlying education for equality and justice include emphasis on co-operation rather than competition and the acceptance of a range of differing and possibly conflicting hypotheses and points of view

- Most primary teachers are not science specialists, and therefore rely on others to provide mutual support in terms of knowledge, ideas and confidence

- Parents can be an important source of support in enabling children to carry out science activities in the environment outside the school, so that collaboration with parents can considerably enhance childrens' interest, understanding and performance in science.

Each of these will be discussed in detail below.

Teacher mediation as an opportunity to 'scaffold' learning.

A group of children have been cutting open fruits to observe their contents, and decided to count the number of seeds in each fruit. They have tried around seven different kinds of fruit. One child is arguing that 'bigger fruits have more seeds', whilst another says, 'no, it depends on what sort it is; if its sour, if it aint got no taste, it'll probably have a lot of seeds'.

As the teacher overhearing this conversation, you have several choices. You could do nothing, and leave them to sort out their disagreement through discussion or by further observations, though at the risk of a stalemate, frustration or loss of interest. You might provide them with one sweet and one sour/tasteless example of the same fruit, to see if they could use these to test the second idea. You might discuss with them the example of a large avocado or mango, which has only one seed. Or you might intervene to structure a discussion starting from the question, 'OK you've each got an idea or hypothesis: how could you test it?'.

Which you choose is going to depend on the context, the fruits available and your knowledge of the children. But all except the first alternative involve you in establishing collaboration between yourself and the pupils, in the form of a structured conversation. Structure is provided by what you give them in terms of materials, and what you say to them, probably in the form of questions. Much classroom science learning depends crucially on this form of mediation.

It will be enhanced where you can maximise the time for listening and discussing with a small group, which implies that other children have to be deprived temporarily of your attention. Critics of group work have pointed out that such methods mean that many children are likely to be unproductively occupied for significant amounts of time: the only way to overcome this is through immaculate organisation of group and other activities in such a way that those children who are not benefiting from your attention at a given time are occupied with activities that enhance learning rather than simply keep them busy. This is far from easy, but it can certainly be done; and it does not necessarily imply that all groups are doing science.

The alternative is to attempt the same kind of collaborative structured conversation with the whole class. There are many times when this is possible; for instance, when discussing a representation (in diagram form) on the board, or when watching a demonstration (for example, using the pedalling of an upturned bicycle as an analogy for 'current' in electricity). However, the drawbacks of this approach are that you can only ever access the ideas of a small proportion of the class at such a time. Research into teachers' science knowledge shows that those teachers with less well-developed knowledge tend to 'close down' discussion sooner, either in groups or with the whole class; whilst those who are confident in their knowledge sustain the open-ended conversation longer, allowing greater collaboration in the development of children's own ideas, as opposed to the memorisation of teacher's definitive statements.

The debate about whole class versus group teaching has been re-opened by the OFSTED sponsored report 'Worlds Apart?' (Reynolds, 1996) which reviewed international surveys of science and maths achievement and attempted to relate these to teaching approaches in different cultures. In the report, questions are asked about the role of cultural factors (attitudes to effort, parental aspirations, status of teachers), systemic factors (amount of time in school, concentration on limited goals, repeating years), school factors (mixed ability grouping, use of specialist teachers, teacher collaboration, frequency of testing and monitoring) and classroom factors (use of textbooks, whole-class interaction, daily routine, narrow ability range).

Contrary to what various commentators in the press and television have said, the report did not advocate specific changes, but recommended that 'the risk in looking outward and trying new practices is worth taking' (Reynolds, 1996, p. 59). What the report did emphasise was the uniqueness of the English primary school situation, in four key respects:

- the wide range of attainment in any given classroom
- the use of complex group-teaching strategies which are onerous to manage
- the complicated nature of the teacher's role in terms of planning, generating materials rather than using textbooks, and
- the multiplicity of goals (academic, social, cultural, behavioural).

Collaborative learning involving teacher mediation is clearly not threatened by the notion that other practices such as use of textbooks and a degree of whole-class teaching are used. Collaboration between teacher and learner is ultimately about effective scaffolding of children's own learning. In science, this can take place during practical group activity, on a 1:1 or teacher:group basis: it can equally well take place with a whole class during that much-overlooked 'rounding off' time after an activity is complete. Frequently, this time is given over to 'writing up' or finishing off; but the 'real' learning, the making sense of evidence, seeing patterns, making inferences, framing conclusions or hypotheses, often takes place at this time, if you manage to focus your children's attention on it. It is at this time that the 'brain experiments' rather than the hands-on work, goes on. Science is fundamentally about ideas, not things; and about testing those ideas against evidence. It is in thinking aloud, sharing, being challenged, that those ideas are developed.

Clearly, then, it is valuable, as well as reasonable, to look at what other countries are doing in science, as touched on in an earlier chapter. But the fact that some countries achieve better results in primary schools with less emphasis on investigative science should not be taken at face value. We need to know more about the tests (did they test largely recall of knowledge, or use of skills?) and about children's subsequent performance in secondary and tertiary education. The evidence of the Reynolds study indicates that in England, for example, though we do relatively poorly at the primary stages, we outperform most other countries in science at higher (i.e. pre-university) levels, and this may have something to do with the way we and others teach at earlier stages.

I have no doubt, though, having seen science taught in primary schools in many countries, that teachers in any country have much to learn from the beliefs and practices of teachers elsewhere. Recently, 8 of my second-year trainees spent 5 weeks teaching in township primary schools in Kwazulu, South Africa, working in classes of 65-110 children, sitting in rows, with no science materials... yet they all came back with new insights into how to manage science more effectively. The great danger is to assume that our way is somehow right: an equal danger is to assume that change will automatically make things better.

Collaborative investigations in science.

'On some occasions, the whole process of investigating an idea should be carried out by the pupils themselves.' (DfEE, 1995, p. 44)

One of the aspects of science which has been shown to present real difficulties to teachers has been this expectation, quoted from the National Curriculum for Science at Key Stage 2. What it means in more detail is that children should be able to:

- make their own suggestions about how to find things out (level 2)

- carry out a fair test, explain why it is fair and provide explanations for observations (level 3)

- recognise the need for fair tests, take account of observed patterns in drawing conclusions, and relate these conclusions to scientific knowledge and understanding (level 4)

- identify the key factors they need to consider in planning investigations, select suitable apparatus, make observations of appropriate precision, repeat observations to offer explanations for observed differences, draw conclusions consistent with available evidence (level 5).

These are the skills underlying investigative science which are consider to be an essential part of children's learning of science. They are what distinguishes 'being scientific' from simply 'knowing some science'. The fact that these skills are a NC requirement doesn't mean that there is universal agreement about their importance at the primary stage, however: no less an authority than Richard Dawkins has argued, in the recent Reith Lecture, that 'science appreciation' would be adequate for the majority who will not become professionally involved in doing science, by which he means being able to take an interest in science, read popular science journals, and keep up to date with developments in a 'Tomorrow's World' manner.

Nevertheless, it is still clear that being able to bring a scientific approach to everyday problems extends anyone's repertoire of strategies for dealing with things; and the skills described above are those that help with exactly that problem-solving. If you stack plates badly and they fall down and break (my favourite domestic hobby-horse!) then it makes sense to ask why they might be falling, how this relates to ideas about balance, stability, friction, centre of gravity...and then to 'test' another way of stacking (big plates at the bottom, smaller plates at the top, no soup-plates under dinner plates...) to see if it increases the stability of the pile and reduces breakages (and arguments!).

Asking children to develop these skills necessitates asking them to collaborate. For example:

- In making suggestions, offering explanations, relating conclusions, these have to be done to someone else
- Decisions about selecting apparatus, repeating observations, levels of accuracy, are best made by consensus
- Carrying out activities, making precise measurements and recording them simultaneously involves co-operation, especially in areas which involve stop-watches, thermometers or forcemeters
- Sharing ideas for explanations widens the range of possibilities to consider
- Listening to alternative accounts provides new conceptual language to enhance understanding.

Co-operation rather than competition.

The alternative to collaboration is often individual work where each child tries to negotiate with you the teacher, and demands more time than you can spare. The activity can thus become one in which children attempt to identify the 'right' answer, and compete to get it right or be first to finish, 'rightness' of the result becoming more important than the process of learning. There are of course times when the right answer is important; equally, there are many times when the process of arriving at the results is as important as the product. The message that learning is fundamentally a collaborative venture seems to me to be more important than messages about who is right or wrong.

One aspect of learning where this message can be crucial is that of stereotyping. Sorting and Classifying are important skills for handling science data, and children clearly need to practise these. But equality and justice in the classroom demand that we do not use these skills divisively, for example when looking at data to do with the children themselves in aspects of Life and Living Processes. Wyvill (1991) has drawn attention to the common practice of using bar-charts to show the distribution of eye colour, hair colour, height and weight, all of which serve to remind children of differences, and can distress those who perceive themselves to be different or not 'normal'. It also accustoms children at an early age to the sometimes dangerous practice of putting people into stereotypical categories, to simplify description.

A far better approach, Wyvill suggests, is to require children to look for similarities and continuous variation, in height, skin colour or eye colour. There are no simple, predetermined categories in these attributes; to divide children into blue eyes and brown eyes is a gross oversimplification, as all children immediately realise if together they observe each other's eyes. Many shades defy easy description, being grey-green, green-black or whatever. When I first worked in a teacher training college in Africa, I was astonished at the many descriptions of skin colour and texture which the students used unselfconsciously to distinguish one from another; and 'black' and 'white' were never amongst them! Continuous variation implies asking children to arrange any physical attribute within a group in terms of a 'spectrum' of qualities. Try asking your class to line up from darkest hair to lightest hair, and see how difficult they find it! Yet what they do see is that all of them have coloured hair, and that there is an infinite variety of hair colour within their group. Of course, there are non-continuous differences between children, such as their blood-group. But the key question to get children thinking collaboratively might be, 'are we alike in more ways than we are different?'

Group investigative work provides the obvious vehicle for effective collaboration in science; yet as various research projects have pointed out, the arrangement of children in groups in itself does not guarantee collaboration. Much more important is the classroom ethos you establish, and the way in which groups are constituted. Gender and cultural roles can easily dominate the way children operate within a group, unless you address them. One way to do this is to ask children themselves to evaluate the success of their collaboration, by asking question such as:

- who did most of the talking in your group today?
- who did the least talking in your group today?
- who had the best ideas in your group today?
- who was leading the activity?
- who did most of the practical activity?
- were all ideas considered?
- was anyone discouraged from participating?
- who participated in recording and reporting?
- did the group make decisions based on the views of everyone?
- do you like working in this group? If not, why not?

What children can learn from collaboration, as both Cowie and Rudduck (1990) and Bennett and Dunne (1992) have suggested, is to value helping one another, to trust one another more, to work effectively outside normal friendship groups and to recognise that other people bring a range of different ideas and skills which benefit everybody. In a practical science activity, this can emerge through joint progress to a clearly-defined goal, such as solving practical problems by coming up with a testable hypothesis, or through individual contributions to a final product, such as gathering separate sets of data and compiling them, or making parts of a construction and assembling them.

Another element in the success of collaborative practical work is getting the group size right, so that everyone is fruitfully occupied, and thinking of the physical layout of work-spaces, so that children are not frustrated by crowding, excessive noise and competition for resources. But within any such framework, groups still need to learn to co-operate, and setting clear expectations across all classroom work is essential. There will always be anxieties about disruption or domination by individuals, and about the need for clear structure to avert problems over discipline and lack of direction. All these need to be addressed openly with the class during 'de-briefing' sessions, and expectations agreed, so that children see the need and feel responsible for the success of collaboration, rather than having a sense of it being imposed.

Mutual support between teachers.

Most primary teachers are not science specialists; those that are have predominantly biological rather than physical science backgrounds. Most come into training with low confidence in their science ability, and many take up their first posts still feeling anxious about being asked science questions they won't be able to answer. You are not alone!

It has already been noted that science knowledge in itself does not make a good teacher of science; what matters is having a wide repertoire of effective ways of representing this knowledge to children. So for example when children show that they have misconceptions about electricity, how it lights a bulb, how it is 'used up' in the process, it is helpful to be able to use representations of various kinds to model what is happening. Some of these might be the use of analogy; water flowing in a pipe, a stream driving a water wheel, a bicycle wheel being turned by pedalling. Others may be enactive; holding hands in a circle and squeezing hands to feel that something can 'flow' round the circuit without any object actually travelling round it. Others such as cut-away models and diagrams might be symbolic.

How does a teacher extend her repertoire of effective representations beyond those learned in initial training and during in-service courses? One obvious way is through collaborative teaching with those more experienced in science work. There is pressure from many official sources to increase the amount of specialist science teaching, particularly at the top end of KS2: but it would be a mistake to assume that specialising solves this problem. In the short term, there are certainly not going to be enough specialists to go round: and it is wrong to assume that the conceptual difficulties increase as children get older.

Much of children's concept formation in science goes on at a very early age; and it is crucial, if you are to build on very young children's emerging ideas, that you yourself can re-present appropriate concepts in a way which serves to clarify and construct, rather than to confuse. Research from many countries suggest that teachers' own misconceptions are strongly resistant to change!

The role of the Science Co-ordinator is partly to help colleagues develop their confidence and skills, and one way to do this is to widen, by sharing, the repertoire of ways of representing science ideas. This sharing can be done through staff development sessions which focus on single-issues; 'how do you teach about... (forces, evaporation, reflection of light...)' and through trying out ideas used successfully by colleagues. It can equally well be done through collaborative teaching in the classroom; by combining classes or spaces, by using shared areas or open plan areas, and by having a more-experienced and less-experienced teacher work together on science activities, so that modelling of techniques and representations can take place. It can be done by acquiring videos of science programmes and reviewing them with colleagues (Clayden and Peacock, 1994).

Extending your repertoire can happen through reading; but even here, discussion of the relevance of a new strategy needs discussing with colleagues (Have you ever tried it? Does it work? Have we got the resources? What happens when...?) I was fortunate that, in my first experience of teaching science with younger children, I worked in a team-teaching situation where we planned and taught collaboratively, and soon learned to have no hang-ups about teaching 'in front of' colleagues. On the contrary: it taught me that learning to work with, and refer to others is beneficial to all. And I also realised that I got a far better and more realistic appraisal of the effectiveness of what I was doing in this way. 'Not knowing the answer' became a matter of willingly referring children to other teachers, as the 'expert' in that specific issue. Security increased enormously, and I learned to value my colleagues far more.

Collaborating with parents

In 1992, several research projects were undertaken at Exeter to investigate parents' perceptions of the new National Curriculum and its Assessment arrangements at Key Stage 1. The general outcome of this series of studies was an awareness that, contrary to what the media frequently told us, the great majority of parents were happy with their children's early schooling, and took a considerable interest in what was happening at the time (Hughes et al., 1994).

In relation to science, however, we found some differences from what was observed in the other core subjects. Teachers found the introduction and assessment of National Curriculum science particularly difficult, being the area in which they were least confident. Initially, many parents were not even aware that their children did science at all in school. At the same time, many parents who themselves had no background in science felt reluctant to ask questions about their children's work and attainment in the subject, even though they showed no such reluctance in relation to English and maths. Thus a situation developed which we characterised in our research as 'Teachers don't tell, parents don't ask'. (Peacock and Boulton, 1995).

This resulted in major discrepancies between teachers' and parents' perceptions. For example, few teachers thought that parents valued or were interested in science; yet the parents themselves were initially very keen to know more. At first, as NC science was being introduced, we were aware that this was interpreted quite negatively by both parents and teachers: parents aren't interested, teachers can't be bothered. However, as the two years leading up to the first assessment passed, there was a change of attitudes; parents became much more understanding of the situation in which teachers had to operate, and teachers began to accept that teachers were interested in finding out about their children's science.

However, when the first reports to parents were issued, we observed that science reports had much less to say about actual attainment than did reports in the other core subjects. Parents and teachers both acknowledged that when parents came in to discuss the reports with teachers, discussion focused more on interest and application than on knowledge and skill attainment. Moreover, there was also an acceptance by parents and teachers that few other opportunities were provided for parents to find out about their children's science work. Those schools which held Science Fairs or sessions on science for parents found them highly successful; but such events were reported by less than 5% of schools in the survey.

Our major concern was that by the end of the study, the majority of parents felt they did not know any more about their children's science than they did at the beginning; at the same time, the proportion of parents wanting to know more had diminished considerably, from around 80% to less than 50%. We got a strong impression that both teachers and parents treated science as less important than English or maths. In fact, most of the parents' knowledge about their children's science came from the children themselves, and not directly from the schools.

Parents clearly wanted to help their children with science, but did not want to 'bother' hard-pressed teachers by making demands. Several recommendations came out of the survey, as follows:

- There was a clear demand from parents for information from teachers about the science their children were going to do in the coming term, rather than about science that was already complete. A simple sheet, to be sent home at the beginning of term, with an outline of the science topics and activities, was valued by many parents, as a means to foster collaboration between home and school. More schools are now doing this

- Ensuring that science attainment was clearly aimed for would have been helped by a more structured report format, touching on the concepts and skills being developed in science

- Displays of work at open days or science fairs were considered very helpful, especially where these were annotated with pointers to what they implied: for example 'this work is an example of children carrying out fair tests' or 'this work demonstrates children's ability to record measurements through diagrams and charts'. Such structured displays were thought to make it easier for parents and teachers to begin a dialogue about children's attainment and progress

- Parents' Evenings which focused on science specifically were thought to be valuable, especially in conjunction with displays. There was a feeling that this would prevent discussion reverting to the 'safe' topics such as behaviour, literacy and numeracy.

- Joint teacher-parent workshops were widely supported by respondents to our survey. Ideas for their format ranged from mixed parent-teacher groups doing science activities together; 'brainstorm' sessions on barriers to communication, or helping with science at home; and sessions where teachers and parents observed pupils being taught and assessed.

- Helping during SATs. Many parents felt they could help by supervising other children while some were being tested (this of course was at the time when practical testing figured more prominently in the SATs) and that they would learn from this involvement. However, few teachers had ever tried to involve parents in this way.

Can teachers and parents collaborate more effectively in children's science learning? Clearly many possibilities exist by which this could happen in theory. Whether or not collaboration takes place depends on the school's ethos, and its ability to make parents feel welcome and valued as contributors to their children's education; and also on the willingness of both teachers and parents to make time, in their already overburdened days, to encourage parental involvement. There was no doubt that in the schools where collaboration was valued, it had a strong impact on everyone's enjoyment and understanding of science.

CHAPTER 8
TAKING A WIDER VIEW OF SCIENCE LEARNING

I have tried throughout this book to take a perspective on science that is wider than the one expected by the National Curriculum for England and Wales; wider than the increasingly prescriptive nature of our teacher education in science (by the time you read this, there will be a National Teacher Training Curriculum as well); and wider than the un-problematic perspective on science that is presented in the popular media. This chapter summarises the reasons why such a perspective is crucial, and what it might mean in practice.

Twenty years ago, after their review of primary education, HMI pointed out with regret that in hardly any primary classrooms did teachers investigate real-life problems, or pursue the interests of children that could be investigated in a scientific way. In the last few years, Claxton (1991) has pointed out that the situation has changed little; the artificialities of 'lab-land' are still more evident than 'real-world' science. To take one small example: when some students of mine were recently asked by teachers to do activities with children in school on separating solids, one asked if she could use iron filings; my response was to ask how often, in real life, did she use or encounter them, and why would anyone need to separate them using a magnet? Sieving flour, salt, rice or pasta would surely make more sense to 9 year-olds. Yet the student had learned from her own schooling that using a magnet to separate iron filings was a 'real' science experiment that they had done in a laboratory in high school! School science,

Claxton says, has taken on a life of its own, separate from real science and from real life.

Like Claxton, Reiss (1993) has illustrated how the science taught in most primary schools is too narrow in its focus, and has become divorced from perennial concerns such as agriculture, health, cooking, clothing, housing, maintenance and work in general. This is hardly surprising, however: the cultural perspective of the History National Curriculum is also so narrow that primary children are not expected to learn anything about our colonial past, and the Geography National Curriculum is so biased towards the UK that not a single country in Africa, South America, Australasia or Asia (except Japan) is a compulsory study area. It is also worth noting that we are the only country in Europe which does not systematically teach a second European language in primary schools. Thus the National Curriculum in general seems likely to condition teachers and children to take a narrow cultural perspective, unless you, as teachers, actively resist this impoverishment.

It is also apparent that children in general start primary school with great curiosity and wonder about the world around them. They are uninhibited in their capacity to observe and hypothesise, to ask 'why...?' and to try to explain what they experience first-hand. They do not compartmentalise this experience, but see every aspect of their environment as a legitimate challenge, to be probed and investigated. They use all their senses, creating new language where necessary to describe and explain their encounters. (Once when picking wild strawberries, my small daughter spotted a particularly large one out of her reach, and called to me 'come and pick it before I un-spot it!') Yet by the time they reach secondary school, it is commonplace to hear teachers and parents bemoan the fact that their children have stopped being curious, are losing interest in science, are discouraged from investigating for themselves, and expect to be given answers. They cease to see a connection between the facts and figures which dominate school science (and which they are mostly expected to memorise and revise), and their own lives and concerns outside the classroom. As a result, many (particularly girls) grow to dislike and reject science. So what might you emphasise if you wished to prevent this, and to retain your children's sense of excitement about investigating their world in a scientific way?

Learning and natural environments

In earlier chapters, I have emphasised the importance of starting from where children are; and nowhere is this more relevant than in their responses to their environment. Ask your children about their earliest recollections. I did this recently with undergraduate students, and almost all described an event outdoors, in a natural environment. Perhaps in an inner-city environment things would be very different. You could find out. Titman (1994) has described different kinds of responses, ranging from the intimate involvement of rural/farming children who are born into a close relationship with weather, animals and plants, through attitudes characterised by affection and ambivalence to complete rejection, such as that of some ethnic minority children in urban environments who do not feel safe outside their homes. Children need opportunities to make sense of the places and spaces they grow up in; science is a unique way to do this, because being scientific simply means testing out their new ideas against evidence, rather than hearsay or old wives tales. (Is it true that if you eat apples or cheese before bed it will stop you sleeping? Does a red sky in the morning actually bring bad weather? Do lots of berries on the holly really come before a severe winter? How do we make sure this stream water is safe to drink?)

However, as Titman shows, children are increasingly constrained by safety concerns, so that they can no longer explore and investigate in the woods, dens, holes, streams, derelict sites and mill-yards that were the everyday playgrounds of children only 25 years ago. They spend more and more time chaperoned by parents, less and less in 'exploring' the things that fascinate them. So in some ways, schools have to provide safe alternatives within their grounds for children to explore, hide, be still and watch; to experience the doing, thinking, feeling and being which generates curiosity, excitement and wonder. You can help by organising your school environment in appropriate ways: Titman's work on 'Learning Through Landscapes' gives extensive advice on how to manage the area around your buildings to provide maximum opportunities for real environmental learning. You can also help by generating new ways of seeing, through encouraging such activities as:

- blind walks (a blindfold walk through a tactile environment, either led by a guide or following a rope)

- mirror walks (holding a mirror at nose level and following a guide, seeing only what is above, rather than what is below, as you walk)

- ant walks (laying a length of string on the ground, lying next to it, and imagining the journey of an ant along the string)

- lying on your back under a tree or in long grass, and seeing it in new ways.

Many more such ideas and activities have been described by Titman (1994) and Cornell (1981).

Every year, we take our first-year undergraduate trainee teachers for a whole day to a forest, to experience the environment in many different ways. Every year it is a highlight of the course, and produces work from the students of far higher quality than anything else they do. One of the activities is a 'blind walk', holding onto a climbing rope which winds its way from tree trunk to tree trunk through bushes and brambles, for about 40 metres down the slope of a hill. I ask each group to explore by touch alone the trees they encounter, and to estimate how many species they found. At the end of the walk, with masks removed, I ask them to look at the same trees and identify them. The response is always fascinating. Everyone has lots to say, in great excitement; everyone feels they have travelled much further! But few can seriously consider the identification question: they guess, but do not look closely at leaves or trunk. It is as though satisfying me by getting the right answer, by whatever means, is all that matters. There is interest, excitement, awareness... but in themselves these were often over-whelmed by what they felt was expected (right answers). This is always their response: knowing it, I now can capitalise on their interest by focusing them on being attentive, making careful observations, comparison, use of keys to distinguish species: the scientific ways of responding which do not come naturally to them.

Another activity which we expect of them in the forest is to find a place sometime during the day when they can sit and be undisturbed for about 15 minutes. We ask them to observe something which catches their attention, follow it, look closely at it in every aspect. Then to sit still, perhaps to close their eyes, and follow the thoughts which come into their heads; to be attentive. On the week following this day in the woods, when presenting her work, one student wrote the following:

> *...strangely enough, nobody teaches us to try to enjoy the dawn or sunset; nobody teaches us to listen to the sounds made by a river, the song of a bird, the murmur of the wind. We are told even less about the art of appreciating flavours or how to look at the night sky and contemplate the stars, or about any of nature's many other manifestations which attune us to her vital rhythms and are the key to achieving human balance...*

This may not be original, and is not simply about science; but for me it was a sign that the day had succeeded in attuning at least one person to the notion of 'wonder' as the starting point for scientific investigation. You might remind me that enjoying the dawn is not part of the National Curriculum; and I might list for you all the many activities you could do in a forest which would cover aspects of the Science National Curriculum. Neither, to me, is relevant to this chapter. Thinking in a wider, global way about learning is about having the courage and confidence to allow anything to be on the agenda for science: and for many children, caring for the environment is definitely on their agenda of important issues. More than that: as I showed in chapter 2, almost every country in the world puts the Natural Environment as a high priority in its elementary science curriculum, though the specific concerns will vary from place to place and from time to time. In 1996, in England for example, we had a partial eclipse of the sun (with a total eclipse due on August 11th 1999, visible in the south-west – don't miss it!); media coverage of the Icelandic volcano which threatened to engulf huge areas of the island and destroy bridges; BSE and the banning of beef from many school meals; the deteriorating quality of water on many of our most popular beaches; the e-coli outbreak in Scotland; animal cruelty and the 'a dog is for life' campaign, alongside '101 Dalmatians'; and continuing controversies over wind-farms and various kinds of renewable energy. These raise questions which are global in significance, and which challenge the very idea of what science and scientists are about.

For example, the publicity given to BSE/CJD has raised many questions, but most of the popular media coverage has failed to question the role of science knowledge in providing answers, and has been much more concerned with the politics of beef exports and slaughtering. It was only in the London Review of Books that I found a journalist willing to question these matters:

> *There are some important questions about BSE and CJD to which there are no answers. Nobody knows – there are theories but no facts – how BSE is transmitted...nobody knows how BSE got into cattle in the first place. Other questions remain unanswered: what victims? What proofs? What agents? What environments cause the hazard? To scientists, the words 'we don't know' are not admissions of helplessness, but starting-points. Science is the art of setting 'don't know' to work.* (Radford, 1996 p.17)

This message is not too difficult for most children, if put in the right language. And it is a crucial point to make about science. So often, media coverage is categorical and final. Journalists begin reports with 'scientists say that...' and go on to imply that science can't be argued with, it has a kind of finality. But this is less and less the case. In a review of a book about the impact of toxic products on us and our environment, the reviewer drew this conclusion:

> *Those concerned to safeguard [our environment] will have to act on information which is less than perfect. In the course of this century, science has penetrated our lives and our environment, yet we are 'flying blind'. It amounts to a great global experiment...we design new technologies and deploy them on an unprecedented scale around the world long before we can begin to fathom their impact on global systems and ourselves. All of us face dilemmas in a world where low-probability, high consequence risks abound. It creates new uncertainties – many of them global in character – which by and large we cannot use past experience to resolve.* (Giddens, 1996 p. 20)

I do not think our children will thank us, as they mature, for not letting them in on this. To me, it is crucial to de-mythologise the power of science to have answers to all our problems, ('all we need is enough money to crack the problem...') And many of the environmental issues are such that science, morality, economics and politics cannot be separated, for example the unequal consumption of the world's energy resources referred to in an earlier chapter. The scientific question which needs asking is: is it possible, technologically and in terms of raw materials, for every family on our planet ultimately to own a car, fridge, TV, computer? What are the resource and pollution implications, for China alone? The moral and political questions are about the right of over-consuming countries to go on doing so at the expense of those less technologically advanced. They will certainly be bigger issues for our children, when they grow up, than they have been for us.

Yet the debate (if there were one!) might at this stage be better focused on what is appropriate for children in the primary phase of their education. So which of the many real-world, global priorities are appropriate to introduce to primary children?

First a list of possible areas to cover, before considering if we should think about them and how.

Health: nutrition, obesity, anorexia, drugs, viruses, sexually transmitted diseases, safety of medicines, 'rich world' and 'poor world' diseases, expensive operations

Food: production, preservation, distribution, influence of seasons and changing climate

Resources: use of forests, rivers, minerals, production of toxic waste, disposal, pollution

Work: making it easier and safer, control technology, robotics, renewable energy, communications, transportation

Housing: construction, materials, insulation, heating, air conditioning, maintenance

Leisure: exercise, fitness, sport, toys, games, use of the senses

Ecology: 'natural economy', agriculture, green issues, our place in the biosphere; the Gaia hypothesis

Science and Religion: cosmology, creation stories, the origins of the universe and life, eternity.

What response does such a list provoke in you? Obviously, you could not discuss all systematically. Many are only tenuously connected to assessed areas of the National Curriculum. Many concern issues about which we ourselves, as teachers, probably need to be much better informed. At the same time, it is difficult to deny that all are important science issues, and that most if not all are likely to arouse curiosity in most children. So should we teach them? Do you teach them? And what do teachers do elsewhere? A few examples might help.

Education about HIV and AIDS is not part of the Science National Curriculum in England and Wales, yet it figures prominently in the primary science curricula of African states such as Botswana, Zimbabwe and Mozambique, where extensive teaching materials have been developed. In these countries, upwards of 15% of children are now born HIV positive; many experience the death of close family from AIDS. The relevance, even to young children, is high. Similarly, the Kenya primary science curriculum puts a high emphasis on health issues, as it does on agriculture and on making work easier, since these also are areas in which most children are in some way engaged from an early age. Boys look after cattle and goats; girls tend crops and use cooking stoves; and all families are dependent on weather and soil fertility for good crops of maize, coffee or sugar cane which yield the surplus to pay for their schooling. Thus the science syllabus requires that they make and improve tools and stoves, investigate soil and weather, identify plant diseases and parasites, find out about traditional remedies and their efficacy, compare these to modern medicine, examine the need for clean water and sanitation and work in the school garden.

The common thread here is relevance and the real-world needs of children. In all these countries, an over-riding priority in primary schools is to pass the leaving examination with sufficient points to qualify for a place in a secondary school, since otherwise primary schooling is in effect the end of their formal education. Sticking to the syllabus and frequent testing are fundamental: teachers and parents are far more concerned to 'teach to the test' than we are. Yet even this has not prevented the curriculum adapting to real-world needs. The con-straints in African states are not those of curricular intentions, but the (low) quality of training for teachers, their (poor) remuneration and (lack of) promotion prospects. In our case, the constraints are somewhat different: teachers are well-trained, reasonably paid and able to progress, especially those with a science background. So what constrains us from tackling the issues listed above?

The political implications of testing, league tables and potential impact on school rolls clearly compels many to go for 'safety first' in deciding on the actual 'curriculum-in-the-classroom'. Media-manipulated scandals about such schools as Culloden have made many teachers wary about being seen as too progressive, especially as this has become a term of abuse amongst government 'experts' on primary schooling. Teachers fear bad Ofsted reports and poor showings in the league tables, because they affect their resources and their intake of pupils. Anything which can be linked to the child-centred Plowden esque philosophy of the sixties is now labelled 'dotty' and has to be stoutly defended, whereas anything which says 'basics' is OK. And there is a tendency to link environmental issues in an equation which goes something like this:

environment = green = left-wing = dotty = not to be trusted

Other legitimate areas for scientific enquiry such as sex education (= promiscuity), nutrition (= veggies and muesli) and a concern for renewable resources (= new-age freaks) can so easily become labelled in this way. And science has always been a fringe subject in the class-conscious hierarchy of our ancient universities, where classics, law and arts were more appropriate for the education of gentlemen. I exaggerate; but not too much. The trouble with science, as Dunbar (1996) says in his book of the same title, is that fewer and fewer bright students are choosing to study it, and fewer and fewer are training to teach it, with the likely effect that the quality of science teaching will deteriorate, and the confidence of teachers to tackle more controversial issues will evaporate. We are already seeing this in secondary schools.

So here is a vicious catch-22. The less well science is taught, the less interested children will be. Their results will deteriorate, and fewer will study it beyond 16, producing less good teachers, who will teach even less well...a downward spiral of mediocrity. And the same phenomenon is apparent in other countries. How do you, as a teacher, break out of it?

First of all you must believe in yourself as a teacher; you must have a vision of education as an exciting process in which you have a contribution to make. (If you simply see yourself as an instrument of current government policy, you would have put this book down long ago!). From here, one way forward is to approach the topics of the National Curriculum from a different perspective. Various materials exist to help you do this, such as the 'Why on Earth...?' materials from Birmingham (Barnfield et. al., 1991). The example reproduced below shows their approach to the topic of Water (fig. 31).

The approach takes account first of the children's own experiences – starting from where they are – and then focuses on establishing the wider world concerns about water. At this stage, children have a say in planning what they want to investigate, and determine a 'web of questions' to explore. Note the author's conclusion that children's questions 'automatically had a global reference'. There is a world of difference, literally, between teaching about bath-time and the goldfish -tank on the one hand, or about preserving rainfall and making dirty water safe to drink on the other. Everyone who has ever gardened or camped knows the importance of the latter, without bringing in the needs of developing countries. Yet acknowledging cultural diversity and the far greater needs of children in poor countries gives these issues a greatly enhanced significance for children. When we are faced with water-shortages, stand-pipes and illness through pollution and water-born diseases (as many tens of thousands in Yorkshire and the South-West have been recently) the idea of relevance strikes more powerfully. Do we always have to wait until a crisis threatens before we consider such things important enough to teach to our children?

And if you don't know enough about these issues; you can read a little, and more valuable still, you can learn alongside your children. Nothing motivates a child to learn more than seeing an adult whom they respect pursuing something out of genuine curiosity to find out. Their learning will deepen alongside yours.

A second approach, as mentioned, is to target attitudes: in other words, to work primarily on children's curiosity, sense of wonder and awareness of environment as a precursor to science investigation of topics they identify. One 'way in', in addition to field-visits already dis-

Figure 31

Developing a global perspective *1. Children planning*

Stage 1. From the children's experience
Our theme on the school plan was 'Water'. I wanted the children to investigate some of the scientific questions and wider issues arising from the theme, whilst also drawing on their own questions and experience. We started off in groups, with a blank sheet; each group had to record what they associated with the word 'water'. This was a deliberately open ended task so that children could bring their own ideas to it.

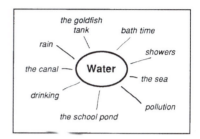

the goldfish tank
bath time
rain
showers
the canal — **Water**
the sea
drinking
pollution
the school pond

Stage 2. From a wider world perspective
As well as building these ideas into the planning, I wanted to make sure that we included a global perspective. To do this, I gave the same group two or three photographs of water use in different parts of Africa. Photographs which had potential for raising issues were chosen. I asked the children to brainstorm questions which they would like to explore in relation to the photographs. Having already drawn on their own experience, they were able to link this in with the photographs and so they asked questions like 'is this water safe to drink?' and 'what do these people do if there isn't any rain?'

Stage 3. What do we want to investigate?
A third stage of the planning was to bring the two earlier stages together and to draw up a web of questions to explore. The children looked at their previous ideas and then made a list of questions which they wanted to look at. It was interesting to see that building in Stage 2 ensured that the questions automatically had a global reference. These are some examples of what children came up with.

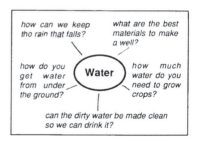

how can we keep the rain that falls?
what are the best materials to make a well?
how do you get water from under the ground?
Water
how much water do you need to grow crops?
can the dirty water be made clean so we can drink it?

cussed, is through the artefacts of artists like Andy Goldsworthy, who uses water, ice, snow, rock, wood, leaves and other materials to create objects of great beauty, utilising many science concepts such as balance, friction, freezing and the colour spectrum in the process (see fig. 32).

This emphasis on attitudes is not necessarily confined to environmental matters. Looking again at the list above, it is clear that children's awareness can be focused on any of the 'big issues' mentioned, be it health, food, work, housing or religious belief. Children are particularly curious about the universe, where it all began, where we came from; yet most know so little even about our own solar system. (Most of my undergraduates, when asked, could not tell or show me where the sun is at noon!). Yet it takes so little to observe the passage of the sun across the sky, or to look at the moon and stars at night. Think of all the questions you would be asked if you could watch the total eclipse with your class in 1999. Sadly, they are likely to be on holiday...but you could prepare for it with them, (e.g. by making pin-hole cameras to observe it) and follow it up when term begins.

You can also record and use some of the vast number of video programmes about these global issues. We have in this country an unequalled output of high quality TV programmes about wildlife, environmental matters, health issues, many of which children will have watched when broadcast but not talked about with anybody. The messages of 'Tomorrow's World' itself are worth discussing and challenging. Richard Dawkins, talking about the public understanding of science, suggested that we need 'science appreciation' like we have 'music appreciation'. If children can discuss pop music and clothes as they do, why not new technology and the science ideas behind them? We simply have to 'come at' these in the right way, i.e. when they are 'in fashion' for some reason, like BSE, or the clockwork radio (how many schools have yet got one?).

It isn't easy to do this. When we feel hard-pressed, most of us take the easy route through our work routines, and most teachers are very hard pressed. I make short cuts in many ways I daren't admit. But I had the good fortune to start my career in West Yorkshire under the guidance of Sir Alec Clegg, and learned the unforgettable message from him, by

Figure 32

his example as well as his doctrine, that teaching, as he put it, is about 'fire-lighting, not pot-filling'. And to light a fire needs a flame, an energy, which in teaching can only come from you, the teacher.

CHAPTER 9

HOW DO WE KNOW WHAT CHILDREN LEARN WHEN THEY ARE TAUGHT SCIENCE?

*We dissect nature along lines laid down by our native languages...
We cut nature up, organise it into concepts, and ascribe
significance as we do, largely because we are parties to an agreement to organise it in this way – an agreement that holds throughout our speech community and is codified in the patterns of our
language... we cannot talk at all except by subscribing to the
organisation and classification of data which the agreement
decrees.* (Whorf, (1956), quoted in Pinker (1995) p. 59).

*The idea that thought is the same thing as language is ...a conven-
tional absurdity if thoughts depended on words, how could a new
word ever be coined? How could a child learn a word to begin
with? How could translation from one language to another be
possible?* (Pinker (1995) p.57).

*Western science teaches students that a rainbow is caused by the
refraction of a beam of light by droplets of water... on the other
hand, traditional [Nigerian] thought explains the appearance of
the rainbow as a python crossing a river Although explaining or
confirming this phenomenon is most difficult, if not impossible, a
non-western student nevertheless holds this as dogma.* (Jegede
(1995) p. 351).

...a pebble is imprisoned
like nothing in the universe.
(Hughes (1995) p.83).

Keep your eye clear
as the bleb of the icicle;,
trust the feel of what nubbed treasure
your hands have known.
(Heaney (1975) p. 20).

Physicists who deal with quantum theory are compelled to use a language taken from ordinary life. We act as if there really were such a thing as a particle because, if we forbade all physicists to speak of particles, they could no longer express their thoughts. (Heisenberg, quoted in Zukav (1979) p. 220).

According to Einstein's ultimate vision, there is no such thing as 'gravity' – gravity is the equivalent of acceleration, which is motion. There is no such thing as 'matter' – matter is a curvature of the space-time continuum. There is not even such a thing as 'energy' – energy equals mass and mass is space-time curvature. (Zukav (1979) p. 199).

In 1922, Werner Heisenberg, as a student, asked his professor and friend-to-be, Niels Bohr, 'If the inner structure of the atom is as closed to descriptive accounts as you say, if we really lack a language for dealing with it, how can we ever hope to understand atoms?'. Bohr hesitated for a moment and then said, 'I think we may yet be able to do so. But in the process, we may have to learn what the word 'understanding' really means. (Zukav (1979) p. 219).

This final chapter is about ideas, language and understanding, in the context of science concepts and our attempts as teachers to develop them in children. The quotations which began the chapter present different, sometimes contradictory, perspectives on this, and will be used to begin to explore how and what we can know about children's learning. Teaching is an active attempt to assist in the practice of learning, or constructing concepts: but as has often been pointed out, there is no guarantee of any connection between what we do as 'teaching' and what is learned by an individual child.

Yet we constantly assess children's learning, sometimes formally, often informally: so what is the problem, you might ask? Doesn't that tell us what they have learned?

Certainly assessment by observation, questioning and tests tells us something. It has become almost a cliché, however, to say that assessing children's science is ultimately about assessing their language. It tells us what children are able to communicate, verbally or graphically or by manipulating things. But how do we know that what a child says is what (s)he understands? Put another way; how far are the concepts 'inside a child's head' equivalent to the words used to describe them? Most of us will admit that, even amongst the most articulate of us, it is often not easy to find words for an idea we have.

> ...*Words strain,*
> *crack and sometimes break, under the burden,*
> *under the tension, slip, slide, perish,*
> *decay with imprecision, will not stay in place,*
> *will not stay still.*
> (Eliot (1944) p. 17).

How much more likely, then, that young children with a limited vocabulary, children who are shy or anxious, children learning in a second language and children from another culture will find difficulty in communicating their ideas to us. And even where they do communicate, we have to ask if their words are to be taken literally, or as unconscious symbols, metaphors or analogies for otherwise inexpressible ideas. Do they do what Heisenberg says physicists do, namely use the language of 'particles' when they know, inside their heads, that matter isn't made up of hard, small billiard-ball-like objects which the word 'particle' connotes?

As an example, I will pursue Jegede's two explanations quoted above, of what makes a rainbow.

Children in different cultures may hold quite different ideas about rainbow-making. Hundreds of research studies have been carried out in recent years, in many countries, to establish the concepts and misconceptions which children hold about many such things. Yet language, worldview and school culture must interact to shape the way children talk. So to what extent is concept development in learners really accessible, or might it be to a greater or lesser degree inaccessible or even incommunicable to others?

Whilst it may be difficult to find out how any child acquired a particular belief about rainbows, it is almost certain that the idea did not come about through first-hand empirical enquiry. Over the past ten years, as I have said, the consensus about children's science learning has been around constructivist theories. The original theory introduced notions such as 'alternative frameworks' and proposed that learners will adapt their explanations (frameworks) to construct a more orthodox conception if they see that a scientific conception is superior, in terms of having greater explanatory power. The theory involves the need for children to articulate their concepts (by speech, writing or drawing) and to modify these in order to develop ideas, and has been refined along various lines referred to as social constructivism (Solomon, 1987, Glasson and Lalik, 1993), radical constructivism (von Glasersfeld, 1991) and more recently contextual constructivism (Cobern, 1993). These variants have developed to take account of the need to acknowledge the mediating role of others such as teachers in a child's construction of concepts, in line with the ideas of Vygotsky in particular. And constructivists have been compelled to consider the environmental context in which learning takes place. Where constructivist research into children's science concepts has cut across cultures, authors have often been eager to point out that common features in children's concepts can be found regardless of culture, for example in Driver (1995).

However, one notable feature absent from most of this research until very recently has been the way in which language influences the concepts as communicated by children. Research has regarded this issue as unproblematic. The predominant tendency until very recently has been for reports of research into children's concepts to say little or nothing about the actual language competency, for example, of the children whose concepts were being investigated. This seems to imply that the authors accept what children say or write is what they understand, regardless of whether this is being communicated in their mother tongue or in a second language, and regardless of any conceptual constraints the language of communication might impose. If a child says that a rainbow is made by a snake crossing a river, should we take it that this is literally the child's idea, without exploring for example the meaning of 'snake' in this context any further? We probably should: yet linguistic concerns such as these raised by Whorf's hypothesis quoted above (that human languages are highly variable and this variability will be reflected in conceptualisation and behaviour) have been notably absent from the constructivist literature. Some recent authors argue that Whorf is quite wrong (as in the quote from Pinker), claiming that there are universal forms of thought which do not require language. Yet even though this is likely to be the case, the issue here is not what children think, but what they communicate to us, whether or not this communication reflects their thoughts, and how it is interpreted by teachers and others (Kawasaki, 1996).

Jegede takes the view that conceptual learning cannot be abstracted from the situations in which it is learned and used: echoing Titman, he suggests that it is the meaning each individual attaches to experiences of the environment which affects learning, and that such meaning is itself the outcome of a complex myriad of socio-cultural factors, one of which is language, a key feature of the teaching-learning environment. Cobern (1996a) points out that the goal of most science curricula is to replace 'traditional' thought with 'scientific thought': and regrets, as I did for different reasons in chapter 8, that school science separates science from pupils' everyday lives, and from their non-school knowledge of the natural world:

> As a science educator, I cannot help but think that there is something awry with the implicit argument that scientific literacy, which all people are said to need, is to be achieved by breaking with the everyday world in which people live and presumably where they will use their scientific literacy. (Cobern, 1996a, p.7).

What ideas, then, do children actually construct? Before confronting this question, it is necessary to examine also the notion of world-views and the part this conception of cultural influence may play alongside language and school culture in the shaping of children's ideas. The notion of world-view has been defined as *a culturally dependent, generally subconscious, fundamental organisation of the mind, which manifests itself as a set of pre-suppostions that predispose one to feel, think and act and react in a certain predictable manner.* (Cobern, 1993).

African traditional world-views, for example, are according to Horton (1971) 'closed systems' in which the possibility of alternative explanations does not exist. Which can be taken to mean, for example, that 'we' in a western scientific culture can consider the possibility of an alternative explanation for the cause of a rainbow, and would then perhaps seek to find ways of testing this; whilst in an African traditional view, the explanation for the rainbow would be so intimately bound up in a belief-system encompassing all aspects of life that it would be inconceivable to consider any alternative explanation without discrediting the whole world-view. Immediately, this creates a division between school science and traditional ways of thinking. Learners may be confronted with having to relate knowledge, presented in an authoritarian way in traditional science classes in a second language, to widely-held yet conflicting belief-systems represented in their mother-tongue.

An illuminating example would be Rendille children, from Northern Kenya. The Rendille believe that God makes all living things and that he (Wakh – a male deity) is alive. Clouds move, give rain and are alive, because rain is the source of life. Air comes from clouds and is therefore alive. Trees are not thought of as alive: but if one is cut down, Wakh can make another. Animals are grouped into only two categories, domestic and wild, all of which are distinguished from other entities by the fact that, like people, they have souls. When a man dies he becomes a 'malaika' or ancestral spirit, part alive, part dead: the Rendille words for spirit, soul and wind are closely related. Malaika live under the ground, and can help or punish people, but not as widely as Wakh is able to do. Wakh is the cause of disease and death (Clarfield, 1987).

However, even such a conflict of Rendille and western-science world-views may not necessarily lead to the driving out of one or other set of ideas and beliefs: Jegede has elaborated the notion of Collateral Learning to describe the process whereby

> *a learner in a non-western classroom constructs, side by side and with minimal interference or interaction, western and traditional meanings of a simple concept... with a capability for strategic use in either the western or traditional environment.* (Jegede, 1995, p. 351)

Note that this idea of collateral learning implies the constructing of a single concept with separate meanings. Yet the language(s) in which the concept and its meanings are constructed is not dealt with in Jegede's article: and I suggest that this is perhaps a crucial aspect of concept construction that is being overlooked. For as Lynch and Jones have pointed out, in the first of several recent studies which confront the role of language in construction of science ideas;

> *the notion of world-view incorporates the constraints of both language and culture on concept learning in an interdependent relationship. Consequently, teachers are dealing with the development, enrichment or change of world-views rather than simply science concepts.* (Lynch and Jones, 1995, p.118).

For constructivism has at its heart the notion that knowledge and belief are not strictly separable; and as Cobern (1996a) makes clear, this helps us to understand how worldview directly influences conceptual development. A similar conclusion was arrived at 70 years ago by a gathering of the world's greatest physicists to confront the dilemmas posed by the then-new quantum mechanics. Their interpretation – a major turningpoint in science – was that whether or not something is 'true' is not a matter of how closely it corresponds to some notion of 'absolute truth', but of how consistent it is with experience. And experience is crucially determined by culture and language.

For example, where a given culture does not have in its language a word for a key concept from another language (such as the lack of a word for 'Nature' in Japanese, or for the supposedly-equivalent word 'Shizen' in English), how do we, confined to our English-language culture, know what concept is constructed by a Japanese student when (s)he is taught about 'nature' in school? (Kawasaki, 1996). That extremely complex myriad of social and cultural factors which influences construction of the meaning of 'Nature' to an English child is inaccessible to the Japanese student, as is the notion of 'Shizen' to you or me, as Kawasaki's list of synonyms will soon make clear. So can we know what the Japanese student has learned?

This is not an academic point, since it is equally applicable to ideas central to science learning within our own National Curriculum, such as the terms Knowledge and Observation, as pointed out by Peat (1995) in his study of science amongst indigenous North Americans. Peat for example describes how amongst the peoples of North America, Knowledge is perceived more as a process than a thing, a verb rather than a noun:

> *Knowledge, to a Native person, cannot be accumulated like money stored in a bank, rather it is an ongoing process better represented by the activity of coming-to-knowing than by a static noun. Each person who grows up in a traditional Native American society must pass through the process of coming-to-knowing, which, in turn, gives him or her access to a certain sort of power, not necessarily power in the personal sense, but in the way a person can come into relationship with the energies and animating spirits of the universe.* (Peat (1995) p. 55).

A similar situation is encountered amongst peoples in various other parts of the world. For example, Lynch (1996) and Lynch and Brown (1995) in their studies of how Tagalog and Ilocano speakers in the Philippines describe the Solar System and the Nature of Matter, observe that they describe such phenomena in terms of their use and relatedness to humans, rather than in terms of their 'thing-ness'. So our idea of 'Knowledge' is dependent on our cultural outlook: there is no universal idea of what 'knowledge' is. When we observe something, we 'see' different things, dependent on what our culture predisposes us to see.

Horton goes on to illustrate how what we in western science call 'theories' are often portrayed in African cultures as 'spirits' or manifestations of some supreme power. While we take the concept of 'chance' or 'chaos' for granted as a sufficient explanation for many physical phenomena, and use mathematics and statistical probability to account for all kinds of behaviour (including, in modern physics, matter itself), in Africa all 'accidents' must be explained, as everything is caused. Spirits have the same function as our theories: Gods are perceived as the 'pointers to a potential theory' (Horton, 1971) providing the link between effects and their causes. Religion plays the role which science plays in our culture: inevitable, then, that children's ideas about phenomena will have different significance.

Thus what we call exploration and investigation – the processes of 'coming-to-knowing', of searching for explanatory theory for phenomena, of abstracting and naming generalisations – seems to go on in all cultures, but with different significance. Yet it seems that we rarely address such matters as what our children understand by these central ideas of the Science National Curriculum itself, such as Investigation, Process, Life, Things, Materials, Physicality. Never mind Einstein's more elusive notions of Gravity, Matter, Mass and Energy, or what a child from another culture (one where there is no practical investigative work in schools, for example) might understand.

Does the fact that, for example, Chinese has no conditional, inhibit Chinese children in their science lessons from asking 'what if...' questions, and entertaining hypothetical situations? Pinker points out that Wintu speakers do not have tenses; but when they convey knowledge, they use different suffixes to indicate whether their knowledge was learned through direct observation or hearsay: would this influence the way children understand and report their science? We have no way of knowing. And how is this relevant, you are asking, to a child in a science classroom in England?

Let us go back to Jegede's question about the rainbow. If you ask a student 'what makes a rainbow?', (s)he will probably know whether or not (s)he knows (how?!); and will perhaps recall, in the mind's eye, an image of a rainbow previously seen. But is a rainbow something 'out there' to be seen? Does it have a concrete physical reality? We are getting into semantic and metaphysical difficulty already...

The student with the 'English' science world-view, speaking in English, may then use the words 'refraction of light from the sun' to explain what causes the phenomenon, but we do not know if this is 'understood' or simply memorised. (When I asked a sample of around 100 each, of primary children and undergraduates many confused 'refraction' with 'reflection', using them almost synonymously). You might pursue the explanation further by asking how the refracted light reaches the eye, (since it is easily observed that a rainbow is always seen in the opposite direction to the sun, therefore light is not directly transmitted to the eye through the water droplets). This would give us an insight into the student's understanding of the actual event, as opposed to his theoretical understanding of refraction, and would probably lead to discussion using the language of reflection, rays, scattering and other concepts related to light and colour.

Or we could pursue questions about how the student knows that the rainbow is made by something (s)he names refraction, and what causes this bending of light. Eventually, I suggest, whichever line of questioning is pursued, the student will refer to an authority – perhaps a teacher or parent or textbook – as the source of his or her knowing. What (s)he knows, in other words, becomes cast in the language of authority; we reach a level beyond which we can no longer 'get at' what the child knows, except that (s)he knows (remembers?) the authoritative version (the 'right answer').

Now imagine questioning an African student in the same way. He might tell us, as Horton does, that amongst the Kalabari of West Africa there are three major categories of lesser gods, namely ancestors, heroes and water-spirits: and that the water-spirits of the Kalabari are 'like pythons'. Water-spirits are 'forces underpinning all that lies beyond the confines of established social order' (Horton, 1971). Hence when the student tells us that the appearance of a rainbow is explained by 'a python crossing a river', we might now interpret this to mean that a rainbow is caused by a force, which has some particular way of acting on water, for which the notion of a 'python crossing' is a successful analogy, in the socio-cultural context of Kalabari world-view, particularly as the coloration and striation of pythons and rainbows are quite similar. And that this view, like that of the English child, is ultimately a view received from some authority, perhaps an elder or diviner. The child again uses the language of authority to represent their understanding, which is inaccessible by other means.

It would be interesting but impossible to find out how the western student would explain his theory about the rainbow to a non-English-speaking Kalabari, and how the Kalabari student would make sense of it within his world-view. As Cobern (1996b) points out – and this to me is why our discussion of other cultural interpretations is relevant – questions about one's relationship with the natural world, which are fundamental to non-western cultures, are rarely asked in our western scientific cultures.

Thus these apparently irreconcilable alternative explanations of the rainbow are probably closer than we at first suspected, and might well exist collaterally. They beg many other questions about the multi-faceted meanings, for children, of terms like 'light' (for which many alternative conceptions have been found amongst children the world over), 'reflect', 'colour' and (in Kalabari culture) words like 'python' and 'river'. (How do you know what you understand by the word 'light'?)

But beyond them is the possibility that a student with access to both Kalabari and western science world-views and the languages in which they are expressed, might construct a unique and (to us) incommunicable concept of rainbow-making. How do we know that the 'rainbow' we perceive is 'actual' i.e. there as perceived? Is the rainbow in the sky, in the eye, or in the mind? Science, like the quote from Heaney's poem, asks us to trust only the evidence of our senses. Children's constructed concepts, on the other hand, and the feelings inextricably linked with them, like rainbow-making, are unique, important and life-enhancing in ways that no theoretical scientific explanation can account for. In terms of knowing what children actually understand, there may not be a crock of gold waiting for us at the end of the constructivist rainbow. As Niels Bohr said, we may have to re-learn what the word 'understanding' means.

This is not a bleak metaphysical scenario, though! What it means to me, as a teacher, is that above all we have to take our opportunities to find out what children understand; and that doing this is more arduous, yet far more rewarding, than simply asking them what they know. It involves valuing and finding out about the children themselves in their culture, the context of values and beliefs which make sense of what they communicate to us about their science ideas. The opportunities we have to teach science are dependent, for their success, on us taking full advantage of our opportunities to learn, fully, about what our children understand; about what they are. The option is to 'teach' (which usually means 'try to put across') ideas which may not connect with children's existing ideas, beliefs or interests, and which may thus simply be retained, briefly, for the sole purpose of being tested to see if they have been retained. Put this way, to me, there is no option.

References

Alexander R., Rose J. and Woodhead C. (1992) *Curriculum Organisation and Classroom Practice in Primary Schools; a Discussion Paper*. London, Department of Education and Science.

Baez A.V. (1991) Teaching youth about the environmental impact of science and technology. in Husen T. and Keeves J.P. (1991) *Issues in Science Education*. Oxford, Pergamon.

Ball D.L. and Feiman-Nemser S. (1988) 'Using textbooks and teachers' guides: a dilemma for beginning teachers and teacher educators'. *Curriculum Inquiry*, 18 (4).

Barnfield M. et al. (1991) *Why on Earth? An Approach to Science with a Global Dimension*. Birmingham: Development Education Centre.

Bennett S.N. and Dunne E. (1992) *Managing Classroom Groups*. Hemel Hempstead: Simon and Schuster.

Berluti A. (1980 onwards) *Beginning Science* (5 volumes; standard 4-standard 8). Nairobi: Macmillan Kenya.

Biber D. (1991) Oral and literate characteristics of selected primary school texts. Text, 11.

Black P. and Atkin J.M. (1996) *Changing the Subject: Innovations in Science, Mathematics and Technology Education*. London: Routledge.

Browne N. (ed. 1991) *Science and Technology in the Early Years*. Buckingham: Open University Press.

Bruner J. (1960) *The Process of Education*. Cambridge, Mass.: Harvard University Press.

Carre C. (1993) *Subject Matter Knowledge and Teaching Performance in Primary Science*. Unpublished PhD thesis, University of Exeter.

Carre C. and Ovens C. (1994) *Developing Primary Teaching Skills*. London: Routledge.

Clarfield G (1987) *The Rendille Ethnosociology of Persons and Action*. Nairobi: Bureau of Educational Research, Kenyatta University.

Claxton G. (1991) *Educating the Enquiring Mind; the Challenge for School Science.* London: Harvester Wheatsheaf.

Clayden E. and Peacock A. (1994) *Science for Curriculum Leaders.* London: Routledge Education.

Cleghorn A. (1992) 'Primary level science in Kenya; constructing meaning through English and indigenous languages'. *International Journal of Qualitative Studies in Education,* 5 (4).

Cobern WW (1993) Contextual constructivism: the impact of culture on the learning and teaching of science. In Tobin K. (Ed) *The Practice of Constructivism in Science Education.* Washington DC: AAA Press.

Cobern WW (1996a) World view theory and conceptual change in science education. *Science Education* (forthcoming)

Cobern WW (1996b) 'Constructivism and non-western science education research'. *International Journal of Science Education* 18(3), 295-310.

Consortium for Assessment and Testing in Schools (CATS) (1990) Pilot Study of SATs for Key Stage 1; an Evaluation Report. London: Schools Examinations and Assessment Council.

Cornell J.B. (1981) *Sharing Nature with Children.* Watford: Exley.

Cowie H. and Rudduck J. (1990) *Co-operative Groupwork in the Multiethnic Classroom.* Sheffield: BP Education Service.

Crossley J. (1991) 'Grouping for Science'. *Primary Science Review* 17, 8-9.

Cummins J. (1983) Language proficiency and academic achievement, in Oller J.W. (Ed) *Issues in Language Testing Research.* Rowley, Mass.: Newbury House.

Cziko GA (1989) 'Unpredictability and indeterminism in human behaviour: arguments and implications for educational research'. *Educational Researcher* 18(3), 17-25.

Department of Education and Science (1985) *Science 5-16: a Statement of Policy.* London: HMSO.

Department of Education and Science (1988) *Science for ages 5-16.* London: DES.

Department for Education (1995) *Key Stages 1 and 2 of the National Curriculum.* London: HMSO.

Desforges C.W., Mitchell C. and Peacock A. (1992) *Assessment at Key Stage 1 in Devon: Final Report.* Exeter: Exeter University School of Education.

Douglass R. and Fraser-Abder P. (1984) *Primary Science for the Caribbean; A Process Approach.* (7 volumes; standard 1-standard 7). London: Macmillan Caribbean.

Dowling P. (1995) A Sociological Analysis of School Texts. Proceedings of the Annual Conference of the Southern African Association for Research in Mathematics and Science Education, Cape Town.

Driver R. (1995) Constructivist approaches to science teaching. In Steffe LP and Gale J. (eds), *Constructivism in Education.* Hove: Lawrence Erlbaum Associates.

Dunbar R. (1995) *The Trouble with Science*. London: Faber and Faber.

Eliot T.S. (1944) *Four Quartets*. London: Faber and Faber.

Ennever L. and Harlen W. (1972) *With Objectives in Mind; a Guide to Science 5-13*. London: Macdonald Educational.

Francis V. (1996) *Personal communication*.

Galton M. and Williamson J. (1992) *Groupwork in the Primary Classroom*. London: Routledge.

Galton M. and Harlen W. (1990) *Assessing Science in the Primary Classroom; vol. 1, Observing Activities*. London, Paul Chapman.

Geertz C. (1973) *The Interpretation of Culture*. New York: Basic Books.

Giddens A. (1996) 'Why sounding the alarm on chemical contamination is not necessarily alarmist'. *London Review of Books*, 5.9.96, 20-21.

Gilbert S.W. (1989) 'An evaluation of the use of analogy, simile and metaphor in science texts'. *Journal of Research in Science Teaching* 26 (4).

Githinji S. (1992) 'Using the environment for science teaching: a teacher's view from Kenya'. *Perspectives* 45, 105-123.

Glasson GE and Lalik RV (1993) 'Reinterpreting the learning cycle from a social constructivist perspective: a qualitative study of teachers' beliefs and practices'. *Journal of Research in Science Teaching* 30(2), 187-207.

Goldsworthy A. (1994) *Andy Goldsworthy*. London: Viking.

Handspring Trust (1993) *Spider's Place: How to Become a Great Detective*. Johannesburg: Handspring Trust.

Handspring Trust (1994) *How do Teachers use Innovative Primary Science Materials? Report on an Evaluation of 'Spider's Place'*. Johannesburg: Handspring Trust.

Harlen W. (1975) *Science 5-13: a Formative Evaluation*. London: Macmillan.

Heaney S. (1975) *North*. London: Faber and Faber, 20.

He Youzhi (1981) *Weighing an Elephant*. Hong Kong: Hai Feng Publishing.

Holliday W.G. (1984). Learning from Science Texts and Materials: Issues in Science Education. Paper presented at the Annual Meeting of AERA, New Orleans, April 1984.

Horton R (1971) African traditional thought and western science. In Young MFD (Ed.), *Knowledge and Control: New Directions for the Sociology of Education*. London: Collier-Macmillan.

Hughes M., Wykeley F. and Nash T. (1994) *Parents and their Children's Schools*. Oxford: Blackwell.

Hughes T. (1995) *New Selected Poems*. London: Faber and Faber.

Hyltenstam K. and Stroud C. (1993) *Final Report and Recommendations from the Evaluation of Teaching Materials for Lower Primary in Mozambique: (II) Language Issues*. Stockholm: Stockholm Institute of Education.

Initiatives in Primary Science Education (IPSE) (1988) *An Evaluation Report*. Hatfield: Association for Science Education.

Jegede OJ (1995) Collateral Learning and the Eco-cultural Paradigm in Science and Mathematics Education. Proceedings of the Southern African Association for Research in Mathematics and Science Education (SAARMSE) Annual Conference, Cape Town.

Jegede O. and Okebukola P. (1991). 'The relationship between traditional African cosmology and students' acquisition of a process skill'. *International Journal of Science Education* 13 (1) pp. 37-47.

Kawasaki K (1990) 'A hidden conflict between western and traditional concepts of nature in Science education in Japan'. *Bulletin of the School of Education Okayama University*, 83, 203-214.

Kawasaki K. (1996) 'Concepts of Science in Japanese and western education'. *Science and Education* 5(1), 1-20.

Kendall B. (1990) *Start Finding Out: Pupils' Book for Standard 4*. Nairobi: Longman Kenya.

Koulaidis V. and Tsatsaroni A. (1996) 'A pedagogical analysis of Science textbooks: how can we proceed?' *Research in Science Education* 26 (1).

Knamiller G. (1984) 'The struggle for relevance in Science education in developing countries'. *Studies in Science Education* 11, 60-78.

Langhan D. (1993) *The Textbook as a Source of Difficulty in Teaching and Learning: a Final Report of the Threshold 2 Proje*ct. Pretoria: Human Sciences Research Council.

Lewis J. (1991) Science in society- impact on Science education. in Husen T. and Keeves J.P. (1991), *Issues in Science Education.*. Oxford: Pergamon.

Lynch PP and Jones BL (1995) 'Students' alternative frameworks: towards a linguistic and cultural interpretation'. *International Journal of Science Education* 17(1), 107-118.

Lynch PP (1996) 'Students' alternative frameworks: linguistic and cultural interpretations based on a study of a western-tribal continuum'. *International Journal of Science Education* 18(3), 321-332.

Macdonald C.A. (1990a). *School-based Learning Experiences: a Final Report of the Threshold Project* Pretoria: Human Sciences Research Council.

Macdonald C.A. (1990b) *Standard 3 General Science Research 1987-88: a Final Report of the Threshold Project*. Pretoria: Human Sciences Research Council.

Makau B.M. and Somerset H.C.A. (1978). *Primary School Leaving Examinations. Basic Intellectual Skills and Equality: some Evidence from Kenya*. Nairobi: Examinations Section Research Unit, Ministry of Education.

Marx G. (1991) Comments (response to chapter by Lewis J (1991) . In. Husen T. and Keeves J. P. (1991) *Issues in Science Education* Oxford: Pergamon.

Millar R. and Driver R. (1987) 'Beyond Processes'. *Studies in Science Education* 14, pp. 33-62.

Murila B. (1996) Preliminary report on the pilot study of the use of science textbooks in standard 4 classes in Kenya. Unpublished research report.

National Curriculum Council (NCC) (1989) *Curriculum Guidance 2: A Curriculum for All.* York, NCC.

National Curriculum Council (1991) *Science Explorations.* York: NCC.

National Institute for Educational Research (NIER). (1986) *Primary Science Education in Asia and the Pacific.* Tokyo, NIER.

Ntebela M. and Frencken H. (1992) 'Teachers against AIDS' in Botswana. Paper presented at the 8th International Conference on AIDS, Amsterdam, July 1992.

Nuffield Primary Science (1993) *Teachers' Handbook, Teachers' Guides, Pupils' Books.* London: Collins Educational.

Parker L. (1990) 'Teaching about electricity: the teacher's concept and the pupil's learning'. *Education 3-13* October 1990, 13-19.

Peacock A. (1989) 'What parents think about Science in primary schools'. *Primary Science Review* 10, pp. 20-21.

Peacock A. (Ed.) (1991) *Science in Primary Schools; the Multicultural Dimension.* London, Macmillan Education.

Peacock A (1993) 'A global core curriculum for primary Science?' *Primary Science Review* 28, 8-10.

Peacock A. (1995a) 'The use of primary Science schemes with second language learners'. *Primary Science Review* no. 38.

Peacock A. (1995b) 'An agenda for research on text material in primary science for second-language learners of English in developing countries'. *Journal of Multilingual and Multicultural Development* 16 (5), 389-402.

Peacock A (1995c) 'Access to science education for children in rural Africa'. *International Journal of Science Education* 17(2), 149-166.

Peacock A. and Boulton A. (1991) 'Parents' understanding of science at Key stage 1'. *Education 3-13* October 1991, 26-29.

Peacock A. and Boulton A. (1995) 'Teacher-Parent Communication about Science at Key Stage 1'. *Education 3-13*, 23(2), 58-68.

Peacock A. and Perold H. (1995) Helping Primary Teachers Develop New Approaches to Science Teaching: a Strategy for the Evaluation of 'Spider's Place'. Proceedings of the Annual Meeting of the Southern African Association for Research into Mathematics and Science Education (SAARMSE), Cape Town.

Peat FD (1995) *Blackfoot Physics.* London: Fourth Estate.

Perold H. and Bahr M. A. (1993) *National Science Education Project dealing with Preconceptions and Problem-Solving Strategies in Primary Science: Report on the Evaluation of the Materials Comprising the Pilot Programme.* Johannesburg: Handspring Trust.

Pinker S. (1994) *The Language Instinct.* London: Penguin.

Postlethwaite T.N. and Wiley D.E. (1992) *The IEA Study of Science (II): Science Education and Curriculum in 23 Countries.* Oxford: Pergamon.

Primary Associations (1992) *The National Curriculum: Making it Work at Key Stage 2.* Derby: The Primary Associations.

Primary SPACE Project (1990 onwards). *Research Reports* (14 titles, various authors). Liverpool: Liverpool University Press.

Qualter A. (1996) *Differentiated Primary Science.* Buckingham: Open University Press.

Radford T. (1996) *Poor Cow.* London Review of Books, 5.9.96, 17-18.

Reiss M. (1993) *Science Education for a Pluralist Society.* Buckingham: Open University Press.

Reynolds D. (1996) *Worlds Apart? A Review of International Surveys of Educational Achievement Involving England.* London: Ofsted/HMSO.

Rollnick M. S. and Rutherford M. (1990) 'African primary school teachers what ideas do they hold on air and air pressure?' *International Journal of Science Education* 12 (1).

Rosier M.J. and Keeves, J.P. (1991) *The IEA Study of Science – 1: Science Education and Curricula in 23 Countries.* Oxford: Pergamon.

Roth K. J. (1985) *Conceptual Change Learning and Student Processing of Science Texts.* Michigan: Michigan State University Institutute for Research on Teaching.

Schools Curriculum and Assessment Authority (SCAA) (1995) *Consistency of Teacher Assessment: Exemplification of Standards (Science, Key Stages 1 and 2).* London: SCAA/HMSO.

Shulman J. (1987) 'Knowledge and teaching; foundations of the new reform'. *Harvard Educational Review* 57 (1), 1-22.

Solomon J (1987) 'Social influences on the construction of pupils' understanding in science'. *Studies in Science Education* 14, 63-82.

Stannard R. (1994) *Uncle Albert and the Quantum Quest.* London: Faber and Faber.

Summers M. and Mant J. (1992) 'Some primary school teachers' understanding of the earth's place in the universe'. *Evaluation and Research in Education* 6(2-3), 95-111.

Taylor R.M. and Swatton P. (1990) *Assessment Matters: no. 1.* London: Schools Examinations and Assessment Council.

Thorp S. (ed. 1991) *Race, Equality and Science Teaching.* Hatfield: Association for Science Education.

Titman W. (1994) *Special Places, Special People: the Hidden Curriculum of School Grounds.* Godalming: World Wide Fund for Nature.

Vachon M. K. and Haney R.E. (1991) 'A procedure for determining the level of abstraction of Science reading material'. *Journal of Research in Science Teaching* 28 (4), 343-352.

van Rooyen H. (1990) *The Disparity between English as a Subject and English as the Medium of Learning (a Final Report of the Threshold Project).* Pretoria: Human Sciences Research Council.

von Glasersfeld E (1991) Knowing without metaphysics: aspects of the radical constructivist position. In Steier F (Ed.), *Research and Reflexivity* 2, 553-571.

Womack D. (1988) *Developing Mathematical and Scientific Thinking in Young Children*. London: Cassell.

Whorf B. (1956) *Language,Thought and Reality*. Cambridge, Mass: MIT Press.

Wragg E.C, Bennett S.N. and Carre C.G. (1991) 'Primary teachers and the National Curriculum'. *Research Papers in Eduation* 4(3), 17-45.

Wray D. and Lewis M. (1995) 'Extending interactions with non-fiction texts: An EXIT into understanding'. *Reading* 29 (1).

Wyvill B. (1991) Classroom ideas for antiracism through Science in primary education. In Peacock A. (ed.) *Science in Primary Schools: the Multicultural Dimension*. London: Routledge.

Zukav G. (1979) *The Dancing Wu-Li Masters: an Overview of the New Physics*. London: Rider.

Index

Resources 46, 137
Reynolds, D. 91, 117-118
Rollnick, M. 97
Rosier 28
Roth, K.J. 99
Royal College of Physicians 65

Safety 85, 131
SATs 6, 9, 34-35, 37, 41, 89, 92, 127
SCAA 46, 90
Scaffolding 24, 114, 118
School grounds 131-133
Science 5-13 2-3
Science for All 3, 7, 19-23, 29, 48
Science in everyday life 3, 16, 20-22,131
Scientific literacy 12, 147
Scientists 22, 43-44, 49, 81, 133-134, 144, 149
SEAC 41, 46, 90
Second-language learners 95-114
Selection 38
Scotland 9
Shizen, 149
Shulman, J. 95
Skills (see Investigation Skills)
Smoking 65-67
Social Constructivism 23, 146
Solomon, J. 14, 146
Somerset, A. 38
Sorting and classifying 84-85
South Africa 14, 16, 96, 99, 101, 108, 119
SPACE Project 6, 14, 20, 23-24, 43, 89, 100
Special Educational Needs 22, 35, 41, 77
Specialist teaching of science 11-12, 15, 39, 123
Spider's Place materials 99-100, 108
Standard Tests (see SATs)
Stannard, R. 30
Stereotyping 121-122, 137
Strategies for science teaching 14
Subject matter knowledge 11, 13, 105, 113, 123
Summers, M. 6, 88

Teacher Assessment (TA) 34, 89-94
Teacher education in science 14,101, 113-114, 129, 132-133, 137
Teacher Training Agency 13
Teachers 37, 39, 43, 46, 63-65, 73, 82, 90, 93, 96, 100-101, 105-108, 114, 115, 117, 123-125, 138, 142, 146, 149, 153
Technology 25, 40, 134-135
Texts in science 89, 95-114, 118
Thorp, S. 22
Threshold Project 96
Timetabling 54-55
Titman, W. 131-132, 147
Total eclipse 133, 140
Topic work 39, 41
Toys 42
Traffic lights 77-78

UNESCO 5
USA 95

Vachon, M.K 98
van Rooyen, H. 96
Visual literacy 98,104-114
von Glasersfeld, E. 146
Vygotsky 89, 146

Water 17, 83-84, 87-88, 138-139
West Riding of Yorkshire 15, 142
Western science 5, 143-150, 153
Whorf, B. 143, 147
Womack, D. 45
Woodhead, C. 39
Work 135
Worksheets 96-97
World Conference on Education for All 12
World-views 19, 145, 147-153
Wragg, E.C. 6
Wray, D. 114
Wyvill, B. 121

Zimbabwe 136
Zukav, G. 144